# Edexcel GCSE (9–1)
# Maths

## Grade 1–3 Workbook

Chris Pearce

William Collins' dream of knowledge for all began with the publication of his first book in 1819.
A self-educated mill worker, he not only enriched millions of lives, but also founded a flourishing publishing house. Today, staying true to this spirit, Collins books are packed with inspiration, innovation and practical expertise. They place you at the centre of a world of possibility and give you exactly what you need to explore it.

Collins. Freedom to teach.

Published by Collins
An imprint of HarperCollins*Publishers*
The News Building
1 London Bridge Street
London
SE1 9GF

Browse the complete Collins catalogue at
**www.collins.co.uk**

HarperCollins *Publishers*
1st Floor, Watermarque Building, Ringsend Road
Dublin 4, Ireland

10 9 8 7 6 5 4

ISBN 978-0-00-832252-6

British Library Cataloguing-in-Publication Data
A catalogue record for this publication is available from the British Library.

Author: Chris Pearce
Expert reviewer: Trevor Senior
Commissioning editor: Jennifer Hall
In-house editor: Alexandra Wells
Copyeditor: Joanne Crosby
Proof reader: Julie Bond
Answer checker: Peter Batty
Cover designers: The Big Mountain Design & Creative Direction
Cover photographs: tr: Ju.hrozian/Shutterstock, bl: Rodina Olena/Shutterstock
Typesetter and illustrator: Jouve India Private Limited
Production controller: Katharine Willard
Printed and bound by CPI Group (UK) Ltd, Croydon, CR0 4YY

**MIX**
Paper from
responsible sources
FSC
www.fsc.org **FSC™ C007454**

This book is produced from independently certified FSC paper to ensure responsible forest management.

For more information visit:
**www.harpercollins.co.uk/green**

# Contents

# How to use this book

This workbook aims to help you build your confidence and surpass your expectations in your Mathematics GCSE. It gives you plenty of practice, guidance and support in the key topics and main sections that will have the most impact when working towards grades 1–3.

These sections are colour coded: Number, Algebra, Ratio, Proportion and Rates of Change, Geometry and Measures, Probability and Statistics.

## Question grades

You can tell the grade of each question or question part by the colour of its number:

Grade 1 questions are shown as 〔1〕

Grade 2 questions are shown as 〔1〕

Grade 3 questions are shown as 〔1〕

## Use of calculators

Questions when you could use a calculator are marked with a ▦ icon.

## Hint boxes

The 'Hint' boxes provide you with guidance as to how to approach challenging questions.

> Hint:    A **common factor** of 18 and 27 is a factor of **both** numbers.

## Key words

The 'Key words' boxes highlight important terms and mathematical language that you will need to understand for your exam. Definitions are provided in the glossary at the end of the workbook.

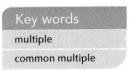

Key words

multiple

common multiple

## Problem solving

This section gives examples of problem solving questions. It helps to build your problem solving and communication skills.

## Checklists

At the end of each chapter, a checklist is provided that lists the key skills and knowledge that have been covered. Use it to identify what you have mastered and what you may still need to work on. By checking off each skill,  you will be putting yourself in the best position to tackle your exam.

## Answers

You will find answers to all the questions in the tear out section at the back of the book. If you are working on your own you can check your answers yourself. If you are working in class your teacher may want to go through the answers with you.

# 1 Multiples, factors and primes

## 1.1 Multiples ✗

**1** Here are some multiples of 5. Write down the next three.

20, 25, 30, 35, 40, _____, _____, _____

**2** Here are the first five multiples of 7. Write down the next three.

7, 14, 21, 28, 35, _____, _____, _____

**3** What is another name for the multiples of 2?

_____

**4** List the first six multiples of 11.

_____

**5** List all the multiples of 20 that are less than 150.

_____

**6** List the multiples of 3 between 20 and 40.

_____

**7** Find a number that is a multiple of 4 **and** a multiple of 5.

_____

**8** Look at these numbers

90    91    92    93    94    95    96    97    98    99    100

a Put a circle ◯ around each multiple of 3.

b Put a square ☐ around each multiple of 4.

c Which number in the list is a common multiple of 3 and 4?

_____

**9**

a  Write down the first six multiples of 4.

_____

b  Write down the first six multiples of 6.

_____

c  Write down two common multiples of 4 and 6.

_____

**10**

a  Write down the first six multiples of 15.

_____

b  Write down the first nine multiples of 10.

_____

c  Write down three common multiples of 15 and 10.

_____

d  Find the **smallest** common multiple of 15 and 10.

_____

**11**  Find **five** common multiples of 2 and 3.

_____

_____

## 1.2 Factors

**1**  $15 = 1 \times 15$    $15 = 3 \times 5$

Use these facts to write down the four factors of 15.

_____

**2**  List all **six** factors of the number 20.

_____

**3**  Two factors of 100 are 2 and 5.

Use this fact to work out two more factors of 100.

_____

_____

1 Multiples, factors and primes

**4**    Here is a list of numbers.    1    2    3    4    5    6

From the list write down the factors of

**a** 8 _____

**b** 12 _____

**c** 15 _____

**d** 17 _____

**5**    **a** Write down the six factors of 18.

_____

_____

**b** Write down the four factors of 27.

_____

_____

**c** Write down the **common factors** of 18 and 27.

_____

> Hint:    A **common factor** of 18 and 27 is a factor of **both** numbers.

**6**    **a** Write down the five factors of 16.

_____

_____

**b** Write down the six factors of 28.

_____

_____

**c** Write down the common factors of 16 and 28.

_____

**7**    Find **three** common factors of 20 and 50.

_____

_____

**8**    121 has three factors. Two of them are 1 and 121. Find the other factor.

_____

# 1.3 Prime numbers

Key word

prime

**1** Here is a list of numbers

10     11     12     13     14     15     16     17     18     19     20

a  Cross out the multiples of 2.

b  Put a circle around the prime numbers.

> Hint:  A **prime** number has just two factors, e.g. 7 and 31 are prime.

**2** a  List all the factors of

i  21 _____          ii  22 _____

iii  23 _____          iv  25 _____

b  Which number in part **a** is prime?

_____

**3** List all the prime numbers between 20 and 30.

_____

_____

**4** Here is a list of numbers     84     85     86     87     88     89     90

Which is the **only** prime number in the list?

_____

**5** Give a reason why each of these numbers is **not** prime.

a  252 _____          b  85 _____          c  999 _____

**6** Write 35 as the product of two prime numbers.

> Hint:   The **product** of 4 and 6 is 24 because $4 \times 6 = 24$.

_____

**7** Write each of the following numbers as the product of two prime numbers.

a  26 = _____          b  33 = _____          c  65 = _____

**8** Work out these products of prime numbers.

a  $2 \times 2 \times 3 =$ _____          b  $5 \times 2 \times 7 =$ _____

c  $2 \times 5 \times 11 =$ _____          d  $5 \times 5 \times 5 =$ _____

**9** Work out the following calculations.

**a** $5 \times 2 \times 2 =$ _____

**b** $5 \times 2 \times 2 \times 7 =$ _____

**10** 28 written as a product of prime numbers is $2 \times 2 \times 7$.

Write each of each of the following numbers as a product of prime numbers.

**a** $18 =$

**b** $77 =$

_____

_____

**c** $30 =$

**d** $32 =$

_____

_____

# 1.4 Lowest common multiple and highest common factor

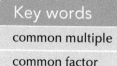

**1** **a** Write down the first six multiples of 6.

_____

**b** Write down the first six multiples of 4.

_____

**c** Write down two common multiples of 6 and 4.

_____

**d** Write down the **lowest** common multiple of 6 and 4.

_____

**2** Find the lowest common multiple of the following:

**a** 2 and 5 _____

**b** 4 and 5 _____

**c** 6 and 5 _____

**d** 10 and 5 _____

**3** Find the lowest common multiple of 2, 3 and 4.

_____

_____

**4** Find the lowest common multiple of 5, 10 and 25.

_____

_____

**5** a List the factors of 12.

_____

b List the factors of 18.

_____

c Find the common factors of 12 and 18.

_____

d Find the **highest** common factors of 12 and 18.

_____

**6** a Find the common factors of 20 and 30.

_____

b Find the highest common factor of 20 and 30.

_____

**7** a Find the highest common factor of 40 and 100.

_____

_____

b Write $\frac{40}{100}$ as simply as possible.

_____

**8** Find the highest common factor of 45, 60 and 75.

_____

_____

_____

_____

1 Multiples, factors and primes

# 1 Problem solving

**1** 60 chairs are put into rows in a hall. There are the same numbers of chairs in each row. All the chairs are used.

   **a** Show that there could be 10 chairs in each row.

_____

_____

   **b** Show that there **cannot** be 8 chairs in each row.

_____

_____

   **c** There is not enough space for more than 14 chairs in each row.

     **i** Find the largest possible number of **rows**.

_____

_____

     **ii** Find the smallest possible number of rows.

_____

**2** 10 can be written as the sum of two prime numbers in two different ways:

$$10 = 3 + 7 \quad \text{or} \quad 10 = 5 + 5$$

   **a** Find **two** ways to write 14 as the sum of two prime numbers.

> Hint: What are the primes that are less than 14?

_____

   **b** Find all the ways to write 24 as the sum of two primes.

_____

_____

   **c** Show that 11 is a prime factor of 385.

> Hint: Divide 385 by 11.

_____

_____

**d** 385 has two other prime factors. What are they?

Hint: Use your answer to part **c**.

_____

**e** 385 has eight factors altogether. You have found three of them. Find the other five.

_____

_____

_____

**Checklist**

I can

☐ list the multiples of a whole number

☐ list the factors of a number

☐ work out the lowest common multiples of two numbers

☐ work out the highest common factor of two numbers

☐ identify prime numbers.

# 2 Fractions, decimals and percentages

## 2.1 Equivalent fractions

**Key words**

equivalent fraction

numerator

improper fraction

mixed number

**1** Write down the fraction shaded. Simplify the fractions as much as possible.

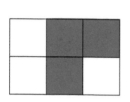
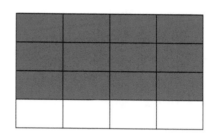
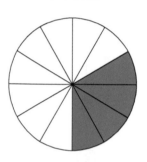

a _____

b _____

c _____

**2** Simplify these fractions as much as possible.

a $\frac{2}{8}$ = _____

b $\frac{4}{10}$ = _____

c $\frac{4}{6}$ = _____

d $\frac{6}{12}$ = _____

**3** Fill in the missing numerators in these equivalent fractions.

a $\frac{1}{2} = \frac{\square}{6}$

b $\frac{3}{5} = \frac{\square}{10}$

c $\frac{1}{2} = \frac{\square}{12}$

d $\frac{6}{8} = \frac{\square}{4}$

e $\frac{3}{2} = \frac{\square}{4}$

f $\frac{2}{3} = \frac{\square}{12}$

g $\frac{3}{4} = \frac{\square}{12}$

h $\frac{5}{2} = \frac{\square}{6}$

**4** Write down three fractions equivalent to $\frac{9}{12}$.

$\frac{9}{12}$ = _____ = _____ = _____

**5** What fraction of £24 are each of the following?

a £12 _____

b £8 _____

c £6 _____

d £18 _____

Write your answers as simply as possible.

**6** Fill in the missing fraction. Write your answer as simply as possible.

**a** 20 kg is _____ of 30 kg

**b** 20p is _____ of 32p

**c** 20 cm is _____ of 50 cm

**d** 60 years is _____ of 80 years

**7** Add these fractions. Write the answer as simply as possible.

**a** $\frac{1}{2} + \frac{1}{2} =$ _____

**b** $\frac{2}{5} + \frac{1}{5} =$ _____

**c** $\frac{3}{8} + \frac{1}{8} =$ _____

**d** $\frac{3}{10} + \frac{1}{10} =$ _____

**e** $\frac{7}{10} + \frac{3}{10} =$ _____

**f** $\frac{5}{8} + \frac{5}{8} =$ _____

**8** Write these fractions as mixed numbers. The first one has been done for you.

**a** $\frac{7}{4} = 1\frac{3}{4}$

**b** $\frac{5}{4} =$ _____

**c** $\frac{9}{4} =$ _____

**d** $\frac{3}{2} =$ _____

**e** $\frac{7}{2} =$ _____

**f** $\frac{11}{8} =$ _____

**g** $\frac{17}{8} =$ _____

**h** $\frac{17}{10} =$ _____

**i** $\frac{33}{10} =$ _____

**9** Write these numbers as improper fractions.

**a** $5\frac{1}{2} =$ _____

**b** $3\frac{1}{4} =$ _____

**c** $2\frac{3}{8} =$ _____

**d** $5\frac{2}{3} =$ _____

**10** Here are four fractions   $\frac{1}{2}$   $\frac{2}{3}$   $\frac{2}{5}$   $\frac{3}{4}$

Put them in order, smallest first

_____

## 2.2 Arithmetic with fractions ✗

**Key word**

denominator

**1** **a** Fill in the missing numerator $\frac{3}{4} = \frac{\square}{8}$

**b** Use your answer to part **a** to work out $\frac{1}{8} + \frac{3}{4} =$ _____

**2** **a** Fill in the missing numerator $\frac{1}{3} = \frac{\square}{9}$

**b** Use your answer to part **a** to work out $\frac{1}{3} + \frac{2}{9} =$ _____

**3** Fill in the missing numerators in the following additions.

**a**

**b** $\frac{1}{8} + \frac{3}{4} = \frac{1}{8} + \frac{\square}{8} = \frac{\square}{8}$

Hint: To add two fractions, make the denominators (bottom numbers) the same.

**c**

**d** $\frac{1}{2} + \frac{1}{10} = \frac{\square}{10} + \frac{1}{10} = \frac{\square}{10} = \frac{\square}{5}$

2 Fractions, decimals and percentages

**4** Work out the following additions:

**a** $\frac{1}{4} + \frac{1}{4} =$ _____

**b** $\frac{1}{8} + \frac{1}{4} =$ _____

**c** $\frac{3}{8} + \frac{1}{4} =$ _____

**d** $\frac{5}{8} + \frac{1}{4} =$ _____

**e** $\frac{1}{10} + \frac{1}{2} =$ _____

**f** $\frac{1}{2} + \frac{3}{10} =$ _____

**5** Do these subtractions. Write the answers as simply as possible.

**a** $\frac{2}{3} - \frac{1}{3} =$ _____

**b** $\frac{4}{5} - \frac{1}{5} =$ _____

**c** $\frac{7}{10} - \frac{2}{10} =$ _____

**d** $\frac{9}{10} - \frac{1}{10} =$ _____

**e** $\frac{7}{8} - \frac{1}{8} =$ _____

**f** $\frac{5}{8} - \frac{3}{8} =$ _____

**6** Work out these subtractions. Write the answers as simply as possible.

**a** $\frac{1}{2} - \frac{1}{4} =$ _____

**b** $\frac{1}{2} - \frac{1}{8} =$ _____

**c** $\frac{5}{8} - \frac{1}{4} =$ _____

**d** $\frac{7}{8} - \frac{1}{2} =$ _____

**e** $\frac{1}{2} - \frac{1}{10} =$ _____

**f** $\frac{1}{2} - \frac{3}{10} =$ _____

**7** Multiply these fractions by whole numbers. The first one has been done for you. Write the answers as simply as possible.

**a** $\frac{2}{3} \times 5 = \frac{10}{3} = 3\frac{1}{3}$

**b** $\frac{1}{2} \times 7 =$ _____

**c** $\frac{1}{4} \times 5 =$ _____

**d** $\frac{3}{4} \times 5 =$ _____

**e** $\frac{1}{3} \times 6 =$ _____

**f** $\frac{4}{5} \times 4 =$ _____

**8** Work these out.

**a** $\frac{1}{3}$ of $7 =$ _____

**b** $\frac{2}{3}$ of $4 =$ _____

**c** $\frac{3}{4}$ of $6 =$ _____

Hint: $\frac{1}{3}$ of 7 is the same as $\frac{1}{3} \times 7$

**d** $\frac{4}{5}$ of $3 =$ _____

**e** $\frac{3}{8}$ of $6 =$ _____

**f** $\frac{7}{10}$ of $5 =$ _____

**9** Complete this multiplication table. Write the answers as simply as possible.

| × | $\frac{1}{4}$ | $\frac{1}{3}$ | $\frac{1}{2}$ | $\frac{2}{3}$ | $\frac{3}{4}$ |
|---|---|---|---|---|---|
| 3 | | 1 | | | |
| 5 | | | $2\frac{1}{2}$ | | |

# 2.3 Fractions and decimals

**1** Write these fractions as decimals.

a $\frac{1}{4}$ = _____ b $\frac{3}{4}$ = _____ c $2\frac{1}{2}$ = _____ d $5\frac{1}{4}$ = _____

**2** Write these fractions as decimals.

a $\frac{3}{10}$ = _____ b $\frac{9}{10}$ = _____ c $\frac{11}{10}$ = _____ d $\frac{27}{10}$ = _____

**3** Write these decimals as fractions. Simplify the answers as much as possible.

a 0.4 = _____ b 0.8 = _____ c 0.6 = _____ d 1.2 = _____

**4** Write these decimals as fractions. Simplify the answers as much as possible.

a 0.15 = _____ b 0.35 = _____ c 0.45 = _____ d 3.55 = _____

**5** Complete this number line by writing the fractions as decimals. The first one has been done for you.

0.125       a _____       b _____       c _____

**6** Here is a number line. Write A, B, C and D as decimals and as mixed numbers.

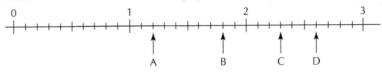

A = _____ = _____          B = _____ = _____

C = _____ = _____          D = _____ = _____

**7** Write these mixed numbers as decimals.

a $8\frac{3}{10}$ = _____ b $4\frac{1}{4}$ = _____ c $6\frac{3}{5}$ = _____ d $5\frac{1}{8}$ = _____

**8** Write these decimals as mixed numbers.

a 9.7 = _____ b 12.8 = _____ c 10.75 = _____ d 8.125 = _____

**9** Write the answers to these multiplications as mixed numbers **and** as decimals.

a $\frac{3}{10} \times 4$ = _____          b $\frac{2}{5} \times 4$ = _____

c $\frac{3}{4} \times 5$ = _____          d $\frac{7}{8} \times 6$ = _____

**10** $\frac{1}{3} = 0.333...$

Write these fractions as decimals in the same way.

**a** $\frac{2}{3} =$ _____

**b** $\frac{1}{9} =$ _____

**c** $\frac{5}{9} =$ _____

**11** Write these numbers in order of size, smallest first.

Hint: First write them as decimals.

$\frac{2}{3}$ $\qquad$ $\frac{4}{5}$ $\qquad$ $\frac{5}{8}$ $\qquad$ $\frac{7}{10}$

smallest _____ largest

# 2.4 Percentages 🖩

Key word

percentage

**1** 37% of the top bar is purple.
10% of the bottom bar is purple.

| 37% | |

| 10% | |

What percentages of each bar are light blue?

_____

**2** Write these percentages as decimals.

**a** 44% = _____

**b** 40% = _____

**c** 4% = _____

**3** Complete the following table.

| Decimal | 0.5 | 0.35 | 0.77 | 0.6 | | |
|---|---|---|---|---|---|---|
| Percentage | | | | | 80% | 6% |

**4** Complete the following table. Write the fractions as simply as possible.

| Percentage | 25% | 70% | 5% | | | 15% |
|---|---|---|---|---|---|---|
| Fraction | | | | $\frac{3}{4}$ | $\frac{9}{10}$ | |

**5** A group of people are in a room. Two-fifths are adult men and 45% are adult women.
The rest are children.
What percentage are children? _____

**6** This is a sum of percentages. 25% + 20% = 45%

Change the percentages to fractions. _____ + _____ = _____

**7** Work out these percentages of £20.

a 25% _____     b 10% _____     c 70% _____

d 75% _____     e 20% _____     f 80% _____

**8** a Write 20% as a fraction. _____

b Work out 20% of

i £15 _____     ii £40 _____     iii £2.50 _____

iv 55 kg _____     v 35 cm _____     vi 80 m _____

**9** Complete the following table.

|       | £10 | £40 | £50 | £80 | £200 |
|-------|-----|-----|-----|-----|------|
| 75%   |     | £30 |     |     |      |
| 40%   |     |     | £20 |     |      |

# 2.5 Working with fractions and percentages

Key words

increase

decrease

**1** Work out

a $\frac{1}{4}$ of 40 kg = _____     b $\frac{2}{3}$ of 18 m = _____

c $\frac{3}{4}$ of 60 people = _____     d $\frac{3}{10}$ of 120 kg = _____

e $\frac{5}{6}$ of 30 g = _____     f $\frac{3}{5}$ of 35 m = _____

**2** a What **fraction** of £40 is £8? _____

Hint: Simplify the fraction $\frac{8}{40}$

b What **percentage** of £40 is £8? _____

**3 a** What fraction of 60 people is

 **i** 45 people _____  **ii** 18 people _____  **iii** 36 people _____

 _____   _____   _____

 **b** Write your answers to part **a** as percentages.

 **i** _____  **ii** _____  **iii** _____

**4 a** Write 32% as a decimal. _____

 **b** Use a calculator to work out 32% of the following.

 **i** £200 _____  **ii** £120 _____  **iii** £25 _____  **iv** £6 _____

**5 a** Write as a decimal

 **i** 8% _____  **ii** 45% _____

 **b** Use a calculator to complete this table.

| | £25 | £40 | £50 | £240 |
|---|---|---|---|---|
| **8%** | | £3.20 | | |
| **45%** | £11.25 | | | |

**6 a** Work out 10% of £60 _____

 **b** Increase £60 by 10% _____

 **c** Decrease £60 by 10% _____

**7 a** Work out 40% of £80 _____

 **b** Increase £80 by 40% _____

 **c** Decrease £80 by 40% _____

**8 a** Work out 5% of 120 kg _____

 **b** Increase 120 kg by 5% _____

 **c** Decrease 120 kg by 5% _____

**9** Jon pays rent of £650 a month. His rent is increased to £676 a month.
Show that this is a 4% increase.

_____

_____

# 2 Problem solving

1. **40** people live in Main Street.

   18 own a bicycle.

   60% take regular exercise.

   $\frac{3}{8}$ are under 18 years old.

   **a** What fraction take regular exercise? Write the answer as simply as possible.

   | Hint: | Start with the percentage you are given. |
   |---|---|

   _____

   **b** How many are under 18? _____

   _____

   **c** What **percentage** own a bicycle? _____

   _____

   90 people live in Back Road. 60% own a bicycle.

   **d** How many **more** bicycles are there in Back Road than there are Main Street? Show your working.

   _____

   _____

2. **a** This is an advert in a shop.

   **i** How much money do you save if you buy two?

   _____

**ii** What is the **percentage** saving if you buy two?

> Hint: Write the saving as a percentage of the full price.

_____

**b** Here is an advert about a sale.

**i** The original price of a coat is £80. Find the sale price. Show your working.

_____

_____

**ii** This is the price ticket on a designer handbag.

Is this a 40% reduction? Give a reason for your answer.

_____

_____

**Checklist**

I can

☐ find equivalent fractions

☐ add and subtract fractions

☐ multiply a fraction by a whole number

☐ convert between fractions, decimals and percentages

☐ work out percentages of a quantity

☐ increase or decrease amounts by a percentage.

# 3 Working with numbers

## 3.1 Order of operations

Key word

bracket

**1** Work out each of the following.

a $3 \times 2 + 5 =$ _____  b $4 \times 4 + 4 =$ _____  c $4 \times 4 \div 2 =$ _____

d $4 \times 4 - 4 =$ _____  e $5 + 3 \times 2 =$ _____  f $4 + 5 \times 5 =$ _____

**2** Work out each of the following.

Hint: Work the brackets out first.

a $2 \times (3 + 3) =$ _____  b $10 \div (3 + 2) =$ _____  c $(3 + 7) \div 2 =$ _____

d $(3 + 7) \times 5 =$ _____  e $5 \times (2 + 3) =$ _____  f $6 \times (3 - 2) =$ _____

**3** Draw a line from each calculation on the left to the correct answer on the right.

a $2 \times 2 + 6$    12      b $2 + 6 \div 3$    2

c $5 - 1 \times 4$    10      d $(3 + 5) \div 4$    5

e $4 \times (5 - 2)$    1      f $20 \div (7 - 3)$    4

**4** Are the following statements true or false? If true, write T. If false, write F followed by the correct answer.

a $12 + 6 \div 3 = 6$ _____      b $14 - 4 \times 2 = 20$ _____

c $16 \div 2 + 2 = 10$ _____      d $8 - 2 \times 2 = 4$ _____

e $2 + 4 \times 3 = 14$ _____      f $6 \times 2 + 8 = 20$ _____

g $21 - 8 \times 2 = 26$ _____      h $(12 - 3) \times 4 = 36$ _____

i $20 \div (5 + 5) = 9$ _____      j $(40 - 20) \div 10 = 38$ _____

**5** James says that $14 - 4 \times 2 = 20$. James is wrong. What was his mistake?

_____

_____

_____

**6** Put in three different numbers to make the following calculation correct.

_____ + _____ × _____ = 17

**7** Work out the following calculations.

**a** $3 \times 3 + 4 \times 4 =$ _____ **b** $(12 - 8) - (5 - 3) =$ _____ **c** $(3 + 7) \times (10 - 4)$

## 3.2 Powers and roots 🖩

**1** $4^2 = 4 \times 4 = 16$

> Hint: Read $4^2$ as 'four squared'.

Write down the following.

**a** $3^2 =$ _____ **b** $7^2 =$ _____ **c** $10^2 =$ _____

**2** Work out the following.

**a** $4^2 + 5^2 =$ _____ **b** $6^2 + 7^2 =$ _____

**3** Work out the following.

**a** $1.3^2 =$ _____ **b** $3.5^2 =$ _____ **c** $4.6^2 =$ _____

**4** $5^2 = 25$, so the **square root** of 25 is 5. We write $\sqrt{25} = 5$.

Write down the following square roots.

**a** $\sqrt{36} =$ _____ **b** $\sqrt{81} =$ _____ **c** $\sqrt{144} =$ _____

**d** $\sqrt{225} =$ _____ **e** $\sqrt{900} =$ _____ **f** $\sqrt{1024} =$ _____

**5** Use a calculator to work out the following.

**a** $\sqrt{1.96} =$ _____ **b** $\sqrt{9.61} =$ _____ **c** $\sqrt{70.56} =$ _____

**6** $4^3 = 4 \times 4 \times 4 = 64$.

> Hint: Read this as '4 cubed'.

Work out the following.

**a** $5^3 =$ _____ **b** $2^3 =$ _____ **c** $6^3 =$ _____ **d** $10^3 =$ _____

**7** Work out the following.

a $1.1^3 =$ _____ b $2.6^3 =$ _____ c $8.1^3 =$ _____

**8** $64 = 4 \times 4 \times 4 = 4^3$

> Hint: 4 is called the **cube root** of 64. We write $\sqrt[3]{64} = 4$

Find the following cube roots.

a $\sqrt[3]{8} =$ _____ b $\sqrt[3]{125} =$ _____ c $\sqrt[3]{1000} =$ _____ d $\sqrt[3]{512} =$ _____

**9** $5^4 = 5 \times 5 \times 5 \times 5 = 625$

> Hint: Read this as '5 to the power 4'.

Work out the following.

a $2^4 =$ _____ b $4^4 =$ _____ c $2^6 =$ _____

**10** Which is larger, $3^4$ or $4^3$? Give a reason for your answer.

_____

_____

**11** Write the following in order of size, smallest first: $4^3$, $3^2$, $2^4$

smallest _____ largest

## 3.3 Rounding numbers 🖩

**1** Write each number to the nearest 10.

a 38 _____ b 682 _____ c 248 _____

**2** Round the following numbers to the nearest 100.

a 427 _____ b 3580 _____ c 3087 _____

**3** Round the following numbers to the nearest 1000.

a 4179 _____ b 5821 _____ c 18 706 _____

**4** Write each number to the nearest whole number.

a 42.7 _____ b 51.9 _____ c 2.4 _____

d 3.5 _____ e 11.8 _____ f 15.1 _____

g 3.23 _____ h 81.22 _____ i 98.61 _____

j 66.53 _____ k 13.45 _____ l 8.99 _____

**5** Circle the value of each number to 1 decimal place. The first one has been done for you.

a 5.24   (5.2)   or   5.3     b 6.75   6.7   or   6.8

c 8.88   8.8   or   8.9     d 14.53   14.5   or   14.6

e 11.621   11.6   or   11.7     f 58.452   58.4   or   58.5

g 8.554   8.5   or   8.6     h 3.751   3.7   or   3.8

**6** Work out the following square roots. Write your answers to 1 decimal place.

a $\sqrt{75}$ = _____ b $\sqrt{130}$ = _____ c $\sqrt{201}$ = _____

**7** Write each of the following to 2 decimal places.

a 4.6666 _____ b 3.34343 _____ c 8.8755 _____

d $\sqrt{5}$ _____ e $\sqrt{52}$ _____ f $\sqrt{90}$ _____

**8** Change the following fractions to decimals. Round your answer to 2 decimal places.

a $\frac{2}{3}$ = _____ b $\frac{1}{6}$ = _____ c $\frac{5}{6}$ = _____

Hint: $\frac{2}{3} = 2 \div 3$

**9** Round the following numbers to 1 significant figure.

Hint: Only one digit in the answer is not 0.

a 821 _____ b 394 _____ c 6129 _____

d 7293 _____ e 4830 _____ f 58 631 _____

## 3.4 Standard form ✖

**1** Write the following as ordinary numbers.

**a** $10^2 =$ _____ **b** $10^3 =$ _____ **c** $10^5 =$ _____

**2** Write the following as powers of ten.

**a** one thousand _____ **b** ten thousand _____ **c** one million _____

**3** Work out the following.

**a** $15 \times 10 =$ _____ **b** $24 \times 100 =$ _____ **c** $38 \times 1000 =$ _____

**4** Work out the following.

**a** $7.3 \times 10 =$ _____ **b** $5.8 \times 100 =$ _____ **c** $2.9 \times 1000 =$ _____

**5** Fill in the following missing numbers. Hint: It will be 10 or 100 or 1000.

**a** $3 \times$ _____ $= 300$ **b** $4.9 \times$ _____ $= 490$ **c** $4.25 \times$ _____ $= 4250$

**6** In **standard form** $6500 = 6.5 \times 10^3$

Hint: Standard form is a number between 1 and 10 times a power of 10.

Write the following numbers in standard form.

**a** $3200 =$ _____ **b** $6000 =$ _____ **c** $1230 =$ _____

**7** Write the following numbers in standard form.

**a** $40\,000 =$ _____ **b** $73\,000 =$ _____ **c** $260\,000 =$ _____

**d** $803\,000 =$ _____ **e** $7\,000\,000 =$ _____ **f** $4\,800\,000 =$ _____

**8** The following numbers are in standard form. Write them as ordinary numbers.

**a** $9 \times 10^2 =$ _____ **b** $9.2 \times 10^2 =$ _____ **c** $3 \times 10^5 =$ _____

**d** $1.8 \times 10^6 =$ _____ **e** $4.75 \times 10^5 =$ _____ **f** $1.41 \times 10^4 =$ _____

**9** $A = 9.3 \times 10^4$      $B = 4.8 \times 10^3$      $C = 3.57 \times 10^4$      $D = 1.101 \times 10^5$

Write A, B, C and D in order of size, smallest first. Hint: Write them as ordinary numbers.

_____

smallest _____ largest

**10** Do the following additions. Give your answers in standard form.

   **a** $2 \times 10^5 + 3.1 \times 10^5 =$ _____

   **b** $2 \times 10^4 + 3.1 \times 10^5 =$ _____

# 3.5 Units

**Key words**

litre

24-hour clock

**1** In the 24-hour clock, 9:30 am is 09:30 and 9:45 pm is 21:45. How many minutes are there between these times?

   **a** 15:15 and 15:40 _____    **b** 07:45 and 08:30 _____

   **c** 09:10 and 10:10 _____    **d** 13:40 and 14:30 _____

**2** Circle the possible height of a woman

   0.165 m      1.65 m      16.5 m      165 m

**3** Circle the possible mass of a man

   0.74 kg      7.4 kg      74 kg      740 kg

**4** Circle the most likely time for an adult to walk 4.5 km.

   45 seconds      4.5 minutes      45 minutes      4.5 hours

**5** Fill in the missing numbers.

   Hint:   1 litre = 1000 ml.

   **a** 2 litres = _____ ml   **b** $1\frac{1}{2}$ litres = _____ ml   **c** $\frac{1}{4}$ litre = _____ ml

**6** Write down a sensible **metric** unit to measure the following.

   **a** The length of your foot _____    **b** The capacity of a glass _____

   **c** The mass of a baby _____    **d** The distance between two towns _____

   **e** The mass of a spoonful of sugar_____    **f** The length of a room _____

**7** Fill in the missing numbers.

   **a** 90 mm = _____ cm             **b** 4 m = _____ cm

   **c** 2 km = _____ m              **d** half a metre = _____ cm

   **e** half a kilometre = _____ m      **f** 5.3 cm = _____ mm

**8** Write the following lengths in metres.

**a** 3.5 km = _____ m        **b** 4.65 km = _____ m        **c** 0.35 km = _____ m

**9** Write the following quantities in ml.

**a** 6.5 litres = _____ ml   **b** 2.85 litres = _____ ml   **c** 0.61 litres = _____ ml

**10** **a** A film starts at 16:40 and lasts 100 minutes. At what time does it finish?

_____

**b** A train journey starts at 09:55 and take 4 hours and 10 minutes. At what time does it end?

_____

# 3 Problem solving

**1** Here are four number cards.

**a** Using two of the cards, fill in the missing numbers to make the following calculation correct.

2 × _____ + 3 × _____ = 22

**b** Using each card once, fill in the missing numbers to make the following calculation correct.

_____ × _____ + _____ × _____ = 23

**c** Using each card once, make the largest possible answer using the following calculation.

_____ × _____ + _____ × _____ = _____

**d** This is a different calculation. Use each card once to make the largest possible answer.

(_____ + _____) × (_____ + _____) = _____

**2** Here are the populations of three towns.

| Town | Grimsby | Blackpool | Sunderland |
|---|---|---|---|
| Population | 88 251 | 126 398 | 287 705 |

**a** Round each population to the nearest thousand. The first one has been done for you.

| Town | Grimsby | Blackpool | Sunderland |
|---|---|---|---|
| Population | 88 000 | | |

**b** Round the population of each town to 1 significant figure.

| Town | Grimsby | Blackpool | Derby |
|---|---|---|---|
| Population | | | |

**c** Find the total population of the three towns. Round your answer to the nearest thousand. Write your answer in standard form.

_____

_____

**Checklist**

I can

☐ carry out operations in the correct order

☐ work out squares, cubes, square roots, cube roots and powers

☐ round numbers to the nearest thousand, to the nearest whole number, to a number of decimal places or to 1 significant figure

☐ understand and use numbers in standard form

☐ convert metric units.

# 4 Sequences of numbers

## 4.1 Describing a sequence 🖩

**Key words**

sequence

term

term-to-term rule

**1** Write down the next number in each sequence.

a 8, 10, 12, 14, _____

b 7, 10, 13, 16, _____

c 10, 14, 18, 22, _____

d 21, 19, 17, 15, _____

e 31, 36, 41, 46, _____

f 56, 52, 48, 44, _____

**2** a Write the next four numbers in the sequence.

13, 16, 19, 22, _____, _____, _____, _____

b State the rule that you used.

_____

**3** Fill in the missing numbers.

a 15, 19, 23, ____, 31, 35

b 44, 48, ____, 56, 60

c 70, 67, ____, 61, 58

d 22, ____, 30, 34, 38

e 50, ____, 62, 68, 74

f 42, 33, ____, 15, 6

**4** Look at this sequence   5, 8, 11, 14, 17, 20, 23, ...   Hint:   5 + 3 = 8; 8 + 3 = 11; 11 + 3 = 14
The **term-to-term rule** is **add 3**.
Write down the term-to-term rule for each sequence.

a 20, 25, 30, 35, ... _____

b 10, 17, 24, 31, ... _____

c 24, 22, 20, 18, ... _____

d 25, 21, 17, 13, ... _____

**5** The first term is 8. The term-to-term rule is **add 5**.

a Write down the next two terms _____

b Work out the 6th term _____

**6**   2, 4, 8, 16, 32, ….

The term-to term rule is **multiply** by 2.

Write down the term-to-term rule for each sequence.

**a**  3, 6, 12, 24, 48, … _____

**b**  2, 6, 18, 54, 162, … _____

**c**  2, 10, 50, 250, 1250, 6250… _____

**d**  1, 4, 16, 64, 256, … _____

**7**   The first term is 5. The term-to-term rule is multiply by 2. Write down the next three terms.

_____

**8**   100, 88, 76, 64, …

**a**  Next term = _____

**b**  Term-to-term rule = _____

# 4.2 Recognising sequences  🖩

| Key words |
| --- |
| square numbers |
| cube numbers |
| triangle numbers |

**1**

$1^2 = 1 \times 1 = 1$     $2^2 = 2 \times 2 = 4$     $3^2 = 3 \times 3 = 9$

1, 4 and 9 are the first three **square numbers**.

**a**  Write down the next three square numbers.

$4^2 =$ _____  $5^2 =$ _____  $6^2 =$ _____

**b**  Write down these square numbers.

$8^2 =$ _____  $10^2 =$ _____

**2**  Circle the square numbers

25       35       49       60       81       110       121

**3**  **a**  Write down the differences between the square numbers.

> Hint:   $4 - 1 = 3$; $9 - 4 = …$

1             4             9             16             25             36

       3        _____        _____        _____        _____

**b**  What do you notice about your answers?

_____

**4**  $1^3 = 1 \times 1 \times 1 = 1$       $2^3 = 2 \times 2 \times 2 = 8$       $3^3 = 3 \times 3 \times 3 = 27$

These are the first three **cube** numbers. Write down the next three.

$4^3 =$ _____       $5^3 =$ _____       $6^3 =$ _____

**5** Show that 1000 is a cube number.

_____

**6**

1          1 + 2 = 3       1 + 2 + 3 = 6

These are the first three **triangle numbers**.

**a** Show that the 4th triangle number is 10.

_____

**b** 5th triangle number = _____

**c** 6th triangle number = _____

**d** 10th triangle number = _____

**7** Show that 64 is both a square number and a cube number.

_____

## 4.3 The $n$th term of a sequence

**1** If $n = 4$ work out the following.

**a** $n + 1 =$ _____ **b** $n + 5 =$ _____ **c** $n - 3 =$ _____ **d** $n + 10 =$ _____

**2** If $n = 3$ work out the following.

**a** $2n =$ _____ **b** $4n =$ _____ **c** $5n =$ _____ **d** $10n =$ _____

**3** If $n = 4$ work out the following.

**a** $2n =$ _____ **b** $2n + 5 =$ _____ **c** $2n - 5 =$ _____ **d** $2(n + 1) =$ _____

**4** Complete the following table.

| $n$ | 1 | 2 | 3 | 4 | 5 |
|---|---|---|---|---|---|
| $2n + 4$ | | 8 | | | |

**5** Complete the following table.

| $n$ | 1 | 2 | 3 | 4 | 5 |
|-----|---|---|---|---|---|
| $2(n + 1)$ | | 6 | | | |

**6** Work out $3n - 2$ when

**a** $n = 1$ _____ **b** $n = 4$ _____ **c** $n = 5$ _____ **d** $n = 10$ _____

**7** The $n$th term of a sequence is $2n + 5$. Write down the following terms.

Hint: 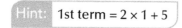 1st term = $2 \times 1 + 5$

1st = _____ 2nd = _____ 3rd = _____ 4th = _____

**8** The $n$th term of a sequence is $3n - 2$. Write down the following terms.

1st = _____ 2nd = _____ 3rd = _____ 4th = _____

**9** Draw a line to join each $n$th term to its sequence.

| $2n - 1$ | 2, 4, 6, 8, 10, … |
|----------|-------------------|
| $2n$ | 1, 3, 5, 7, 9, … |
| $2n + 1$ | 4, 6, 8, 10, 12, … |
| $2(n + 1)$ | 3, 5, 7, 9, 11, … |

**10** The $n$th term is $3n + 5$. Work out the following terms.

**a** 2nd term = _____ **b** 6th term = _____ **c** 10th term = _____

**11** The $n$th term is $10n + 1$.

**a** Write down the first 5 terms

_____

**b** Write down the 10th term

_____

# 4 Problem solving

**1**   **a**  Write down the first five odd numbers.

_____

   **b**  Add the following odd numbers.

    **i**  1st + 2nd = _____  **ii**  2nd + 3rd = _____  **iii**  3rd + 4th = _____  **iv**  4th + 5th = _____

   **c**  Describe the answers to part **b**.

_____

   **d**  Write down the first five triangle numbers.

    Hint:  Remember they are 1, 1 + 2, ...

_____

   **e**  Add the following triangle numbers.

    **i**  1st + 2nd = _____  **ii**  2nd + 3rd = _____  **iii**  3rd + 4th = _____  **iv**  4th + 5th = _____

   **f**  What type of numbers are the answers to part **b**?

_____

   **g**  Add the following triangle numbers.

    **i**  5th + 6th = _____      **ii**  9th + 10th = _____

**2**   These patterns are made from **dots** and **lines**.

1   2   3   4

Pattern 2 has 6 dots.

   **a**  Complete the following table.

| Pattern | 1 | 2 | 3 | 4 |
|---------|---|---|---|---|
| Dots    |   | 6 |   |   |

   **b**  Work out the number of dots in pattern 6.

_____

Pattern 2 has 7 lines.

**c** Complete the following table.

| Pattern | 1 | 2 | 3 | 4 |
|---------|---|---|---|---|
| Lines   |   | 7 |   |   |

**d** Work out the number of lines in pattern 7.

_____

The number of lines in pattern $n$ is $3n + 1$.

**e** Show this formula is correct for pattern 4.

_____

**f** Work out the number of lines for pattern 20.

_____

**Checklist**

I can

☐ work out and use the term-to-term rule for a sequence

☐ identify square numbers, cube numbers and
triangle numbers

☐ use the $n$th term of a sequence to write down terms.

# 5 Coordinates and graphs

## 5.1 Coordinates 🀰

**1**

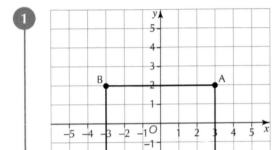

The coordinates of B are (−3, 2).

**a** Write down the coordinates of the other three vertices, A, C and D.

A = _____     C = _____

D = _____

**b** The midpoint of AB is (0, 2). Mark it with a cross.

Hint:   Half way between A and B

**c** Put a cross on the midpoint of each of the other sides. Write down their coordinates.

_____

**2**

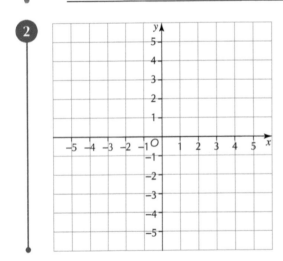

**a** Put these points on the grid. Join them in order with straight lines.

(5, 1)  (5, 3)  (1, 5)  (−3, 3)  (−3, 1)  (1, −1)  (5, 1)

**b** You have drawn a hexagon. Draw the diagonals by joining opposite corners.

**c** Write down the coordinates of the point where the diagonals meet.

_____

**3**

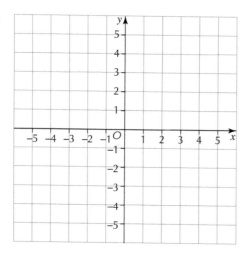

a  Mark each point with a cross. (3, 0) (−1, 2) (5, −1) (−3, 3)

b  Join the points with a straight line.

c  Write down the coordinates of the point where the line crosses the y axis. _____

**4**

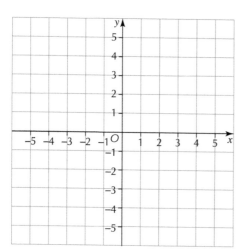

a  Mark **five** points with an x-coordinate of 2.

b  Join them with a straight line.

c  Mark five points with an y-coordinate of −4.

d  Join them with a straight line.

e  Write the coordinates of the point where they cross.

_____

## 5.2 The equation of a straight line

Key word

equation

**1**

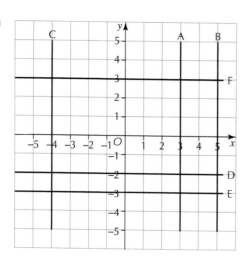

a  The equation of line A is x = 3.
   Write down the equation of
   Line B _____ line C _____

b  The equation of line D is y = −2.
   Write down the equation of

   line E _____

   line F. _____

**2  a** Work out the following.

  **i**  −1 + 3 = _____     **ii**  −2 + 3 = _____     **iii**  −5 + 3 = _____

**b** Complete the following table.

| $x$ | 2 | 0 | −2 | −4 |
|---|---|---|---|---|
| $y = x + 3$ | 5 | | | |

**c**

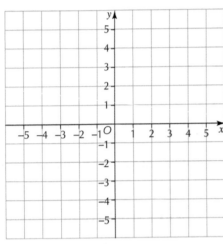

The table gives the coordinates of four points. Plot them on the grid.

**d** Join the points to show the straight line $y = x + 3$.

**e** Write down the coordinates of two more points on the line $y = x + 3$.

_____

**3  a** Work out the following.

  **i**  $2 \times -2$ = _____     **ii**  $2 \times -3$ = _____     **iii**  $2 \times -5$ = _____

**b** Complete the following table.

| $x$ | −2 | −1 | 0 | 1 | 2 |
|---|---|---|---|---|---|
| $y = 2x$ | | −2 | | | |

**c**

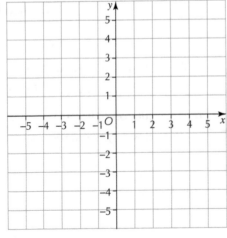

The table gives the coordinates of five points. Plot them on the grid.

**d** Join the points to show the straight line $y = 2x$.

**e** Write down the coordinates of one more point on the line $y = 2x$.

_____

**4** $x + y = 4$

  **a**  **i**  when $x = 1$, $y =$ _____     **ii**  when $x = 3$, $y =$ _____

    **iii**  when $x = 0$, $y =$ _____    **iv**  when $x = 5$, $y =$ _____

  **b**  Complete the following table when $x + y = 4$.

| $x$ | −1 | 0 | 1 | 2 | 3 | 4 | 5 |
|-----|----|---|---|---|---|---|---|
| $y$ |    |   | 3 |   |   | 0 |   |

  **c**  Plot the points on a grid. Join them with a straight line.

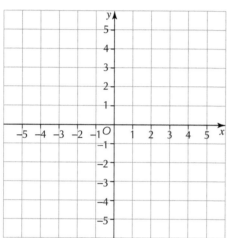

## 5.3 Intercept and gradient ✗

**Key words**

intercept

gradient

**1**  Three points on a straight line are (2, 4), (4, 6) and (−3, −1).

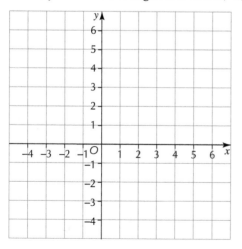

  **a**  Draw the line on the grid.

  **b**  The **intercepts** are the points where the line crosses the axes.

    Write down their coordinates.

    _____

**2** The intercepts of a straight line are (6, 0) and (0, 3). Draw the line on the grid.

**a**

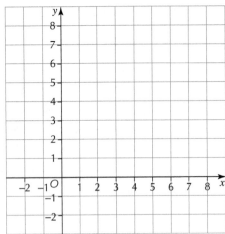

**b** Circle the coordinates of points that are on the line.

(1, 4)   (4, 1)   (8, −1)   (−1, 8)   (3, 2)

**3** Circle the correct answer.

**a** The gradient of line 1 is   1   2   4   8

**b** The gradient of line 2 is   $\frac{1}{9}$   $\frac{1}{3}$   3   9

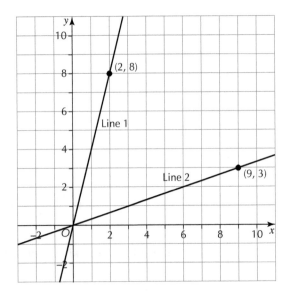

**4** Work out the gradient of the following lines.

**a** OA _____

**b** OB _____

**c** OC _____

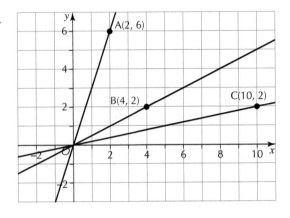

**5** The equation of a straight line is $y = 2x + 4$.

**a** Complete the following table.

| $x$ | −2 | −1 | 0 | 1 | 2 | 3 |
|-----|-----|-----|-----|-----|-----|-----|
| $y = 2x + 4$ | 0 | 2 | | | | |

**b** Draw the line on the following grid.

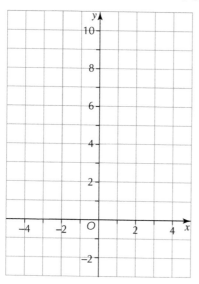

**c** Write down

   **i** the intercept on the $y$-axis = _____    **ii** the gradient. _____

# 5.4 Quadratic graphs

**Key word**

quadratic

**1** **a** Complete the following table.

Hint:   $4^2$ is $4 \times 4$

| $x$ | 0 | 1 | 2 | 3 | 4 | 5 |
|-----|-----|-----|-----|-----|-----|-----|
| $y = x^2$ | 0 | 1 | | | 16 | |

**b** Plot the points on this grid.

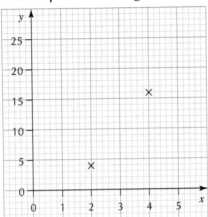

Hint: The scales on the axes are different.

**c** Join the points to draw the graph $y = x^2$.

Hint: Join the points with a smooth curve.

**2**  **a** Complete the following table.

| $x$ | 0 | 1 | 2 | 3 | 4 |
|---|---|---|---|---|---|
| $y = x^2 + 4$ | 4 | | 8 | | |

**b** Plot the points on the following grid.

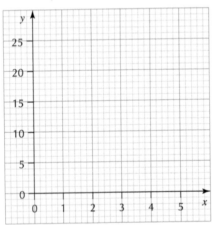

**c** Join the points to draw the graph $y = x^2 + 4$.

**3**  **a** Complete the following table.

| $x$ | 0 | 1 | 2 | 3 | 4 | 5 |
|---|---|---|---|---|---|---|
| $y = x^2 - 10$ | −10 | | −6 | | | |

**b** Draw the graph $y = x^2 - 10$.

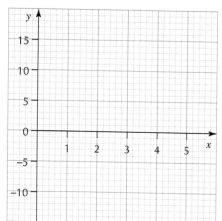

**4** **a** Work out the following.

> Hint: The square of a negative number is positive.

    **i** $(-2)^2 =$ _____

    **ii** $(-4)^2 =$ _____

**b** Complete the following table.

| $x$ | $-4$ | $-3$ | $-2$ | $-1$ | 0 | 1 | 2 | 3 | 4 |
|---|---|---|---|---|---|---|---|---|---|
| $y = x^2$ | | | | | | | | | |

**c** Draw the graph $y = x^2$ on this grid.

> Hint: The $y$-axis is a line of symmetry.

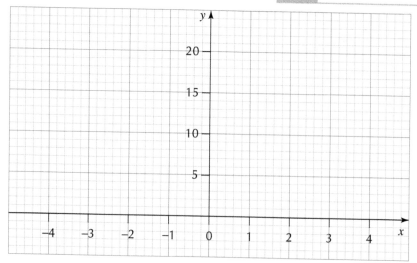

# 5 Problem solving

**1**

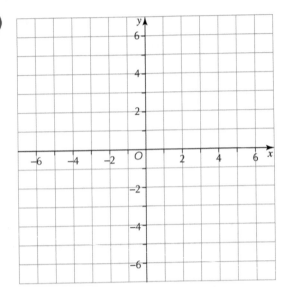

**a** Plot A(6, 3), B(−4, 3) and C(6, −3).

**b** Write down the coordinates of the midpoint of AB. _____

Hint: Draw AB.

**c** Write down the coordinates of the midpoint of AC. _____

**d** Work out the equation of the straight line through A and B. _____

**e** Draw the line $y = x$

**f** Write down the coordinates of the point where AB and $y = x$ cross. _____

**g** A, B and C are three vertices of a rectangle.
Write down the coordinates of the fourth vertex. _____

**2**

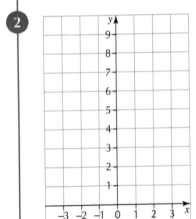

**a** Complete the following table.

| x | 0 | 1 | 2 | 3 |
|---|---|---|---|---|
| 3x | | | 6 | |

**b** Draw the line $y = 3x$

**c** Write down the gradient of the line $y = 3x$. _____

**d** Complete the following table.

| x | −3 | −2 | −1 | 0 | 1 | 2 | 3 |
|---|---|---|---|---|---|---|---|
| x + 6 | 3 | | | | | 8 | |

**e** Draw the line $y = x + 6$.

**f** Write down the gradient of the line $y = x + 6$. _____

**g** Write down the intercept on the y-axis of $y = x + 6$. _____

**h** Complete the following table.

| x | −3 | −2 | −1 | 0 | 1 | 2 | 3 |
|---|---|---|---|---|---|---|---|
| x² | | 4 | | | | 4 | |

**i** Draw the curve $y = x^2$

**j** Write down the coordinates of the point where the three lines meet. _____

Checklist

I can

- ☐ plot points with positive or negative coordinates
- ☐ draw a straight line from its equation
- ☐ find the gradient of a straight line
- ☐ find the intercepts of a straight line, i.e. where it crosses the axes
- ☐ draw a simple quadratic curve.

# 6 Expressions and equations

## 6.1 Substituting into formulae ✗

**Key words**

expression

formula

substitute

**1** Rex is $R$ years old.

Write an expression using $R$ to write these ages. The first one is done for you.

a Keri is 5 years younger than Rex. $R - 5$

b Lucy is 4 years older than Rex. _____

c Tom is twice the age of Rex. _____

d Jose is half the age of Rex. _____

**2** If $w = 6$, what is the value of these expressions?

a $w + 9 =$ _____    b $w - 5 =$ _____    c $w \div 2 =$ _____

d $w - 8 =$ _____    e $2w =$ _____    f $5w =$ _____

**3** $p = 4$ and $q = 6$. Find the values of the following expressions. The first one is done for you.

a $5p + 2q = 20 + 12 = 32$

b $5p - 2q =$ _____ $=$ _____

c $10p + 3q =$ _____ $=$ _____

d $\frac{1}{2}p + \frac{1}{2}q =$ _____ $=$ _____

e $2(p + q) =$ _____ $=$ _____

f $5(q - p) =$ _____ $=$ _____

**4** A formula for estimating the perimeter of a circular pond is

perimeter = 3 × diameter

a Work out an estimate of the perimeter of a pond with diameter 4 m.

_____

b Work out an estimate of the perimeter of a pond with diameter 6 m.

_____

**5** The formula for the perimeter of a triangle with sides of length $a$, $b$, and $c$ is

$P = a + b + c$

a What is the perimeter of the triangle when $a = 3$ cm, $b = 4$ cm and $c = 5$ cm?

_____

b What is the perimeter of the triangle when $a = 12$ m, $b = 6$ m and $c = 9$ m?

_____

**6** The cost £C of hiring a car for D days is given by the formula

$$C = 20D + 10$$

Work out the cost for

**a** 2 days _____ **b** 4 days _____.

**7**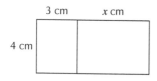

3 cm     x cm

4 cm

The area of this rectangle is $4(x + 3)$ cm².

Work out the area when

**a** $x = 2$ cm _____ **b** $x = 17$ cm _____.

# 6.2 Simplifying expressions

**1** Simplify as much as possible.

**a** $x + 4x = $ _____ **b** $3y - y = $ _____

**c** $p + 2p + 5p = $ _____ **d** $10t - 3t - 2t = $ _____

**2** Simplify the following expressions as much as possible.

**a** $t + 1 + t + 2 = $ _____ **b** $2x + 1 + 2x - 1 = $ _____

**c** $4 + 4k - 2 - k = $ _____ **d** $n + 1 + 3n - 4 = $ _____

**3**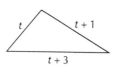

Write an expression for the perimeter of this triangle as simply as possible.

_____

**4** Multiply out the following brackets.

**a** $2(x + 1) = $ _____ **b** $3(y - 2) = $ _____

**c** $4(a + 3) = $ _____ **d** $5(2a - 1) = $ _____

**5** Multiply out the following brackets.

**a** $4(x + y) = $ _____ **b** $3(a + 2b) = $ _____

**6** Draw a line between the following equivalent expressions. One has been done for you.

| | |
|---|---|
| $4(x + 2)$ | $8x + 2$ |
| $2(x + 4)$ | $8x + 4$ |
| $8(x + 1)$ ——————— $8x + 8$ | |
| $2(4x + 1)$ | $2x + 8$ |
| $4(2x + 1)$ | $4x + 8$ |

**7** Factorise the following expressions.

**a** $2c + 10 =$ _____

**b** $3x - 9 =$ _____

**c** $4w + 10 =$ _____

**d** $8n - 2 =$ _____

**8** Factorise the following expressions fully.

**a** $4a + 4b =$ _____

**b** $6x - 12y =$ _____

**c** $20s + 30t =$ _____

**d** $4a + 8b - 12c =$ _____

## 6.3 Solving equations

**Key words**

equation

solve

**1** Solve the following equations.

Hint: Find the answers by subtraction.

**a** $x + 5 = 20$     $x =$ _____

**b** $y + 12 = 19$     $y =$ _____

**2** Solve the following equations.

**a** $2t = 8$     $t =$ _____

**b** $4y = 24$     $y =$ _____

**3** Solve the following equations.

Hint: The answer to part **a** is **not** 3.

**a** $g - 4 = 7$     $g =$ _____

**b** $r - 2 = 10$     $r =$ _____

**4** Solve the following equations.

Hint: Find each answer by doing a multiplication.

**a** $\frac{1}{2}p = 3$     $p =$ _____

**b** $\frac{T}{2} = 9$     $T =$ _____

**5** Solve the following equations.

**a** $2(x + 3) = 14$

**b** $3(y - 2) = 18$

**6** Solve the following equations.

**a** $3y + 5 = 17$

**b** $4x - 3 = 29$

**c** $5w + 12 = 47$

**d** $12g - 32 = 16$

**e** $\dfrac{m}{3} - 8 = 2$

# 6 Problem solving

1  This is a square. Each side has length $L$ cm.

   **a** Show that the perimeter of the square is $4L$ cm.

   _____

   **b** Here are two squares put together.
       Write an expression for the perimeter in centimetres.
       Write your answer as simply as possible.

   _____

   **c** This is a rectangle. The sides are $L$ cm and 3 cm.

   This shape is made from two squares
   (in part **a**) and two rectangles.

   Write an expression for the perimeter. Write your answer as simply as possible.

   _____

2  **a**

   $\boxed{x = 1}$ → $\boxed{x + 5 = 6}$ → $\boxed{2(x + 5) =}$

   Write the missing number in the last box.

   **b**

   $\boxed{x = 4}$ → $\boxed{x + 5 =}$ → $\boxed{2(x + 5) =}$

   Write in the two missing numbers.

   **c**

   $\boxed{x =}$ → $\boxed{x + 5 = 12}$ → $\boxed{2(x + 5) =}$

   Write in the two missing numbers.

   **d**

   $\boxed{x =}$ → $\boxed{x + 5 =}$ → $\boxed{2(x + 5) = 40}$

   Write in the two missing numbers.

# 7 Ratio and proportion

## 7.1 Ratio notation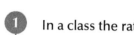

**Key word**

ratio

**1** In a class the ratio of boys to girls is 2 : 1. There are 8 girls. Work out the number of boys.

_____

**2** A hockey team has seven women and four men.

**a** Write down the ratio of women to men.

_____

**b** Write down the ratio of men to women.

_____

**3** A class has 20 girls and 10 boys. Write down the ratio of girls to boys. Give your ratio as simply as possible.

_____

**4** There are 12 cats and 4 dogs. Complete these sentences.

**a** There are _____ cats for every dog.

**b** The ratio of cats to dogs is _____ : 1

**5** Write these ratios as simply as possible. The first one has been done for you.

**a** 8 : 4 = 2 : 1     Hint: Divide both numbers by 4.

**b** 9 : 3 = _____        **c** 20 : 5 = _____

**d** 6 : 8 = _____        **e** 18 : 12 = _____

**f** 10 : 25 = _____        **g** 16 : 24 = _____

**6** Write down the ratio of red circles to blue circles in the following patterns. Write the ratio as simply as possible.

**a**

_____

**b**

_____

**c**

_____

**7** Here are the ingredients for a recipe.

Write the ratio of the masses of each of the following. Give your answers in simplest form.

**a** tomatoes : cheese = _____

**b** pasta : cheese = _____

*Recipe*

Pasta          400 g

Tomatoes   500 g

Cheese       250 g

**8** Write down the following ratios for one person as simply as possible.

**a** Noses to toes. _____

**b** Fingers to thumbs. _____

**9** The ratio of cows to sheep in a field is 3 : 2.
Write whether these statements are **true** or **false**.

**a** There are more cows than sheep. _____

**b** Over half the animals are sheep. _____

**c** The ratio of legs to animals is 4 : 1. _____

**d** 3 animals out of 5 are cows. _____

**10** In a group of people one quarter are men and the rest are women.
Work out the ratio of women to men.

_____

**11** $\frac{3}{5}$ of the plates in a cupboard are large. The rest are small.
Work out the ratio of large plates to small plates.

_____

# 7.2 Dividing using a ratio

**1** There are 15 children in a room.

One-third are boys and the rest are girls. Work out

**a** the number of boys

_____

**b** the number of girls

_____

**c** the ratio of boys to girls. Write your ratio in its simplest from.

_____

**2** There are 40 vehicles in a car park.

$\frac{3}{4}$ are cars and the rest are vans.

Work out

**a** the number of cars _____

**b** the ratio of cars to vans. Write your ratio in its simplest from.

_____

**3** The ratio of dry days to wet days in June is 5 : 1.

There are 30 days in June.

Work out the number of

**a** wet days _____     **b** dry days. _____

**4** The ratio of apples to oranges in a bowl is 1 : 3.

Work out

**a** the number of oranges if there are 4 apples

_____

**b** the number of apples if there are 18 oranges.

_____

**5** 60 people take a driving test. 10 people fail. The rest pass.

Work out the ratio pass : fail. Write your ratio in its simplest form.

_____

**6** The ratio of girls to boys is 2 : 3.

   **a** If there are 20 children in total, work out the number of girls.

_____

   **b** If there are 20 girls, work out the number of boys.

_____

**7** In a recipe the ratio of the mass of flour to the mass of butter is 4 : 1.

   **a** If the total mass is 250 g, work out the mass of butter.

_____

   **b** If the mass of flour is 80 g, work out the mass of butter.

_____

# 7.3 Proportion 🖩

**Key words**

proportion

speed

scale

**1** Sam walks 6 km in 1 hour.

Assume that she walks at a constant speed.

Work out

   **a** the distance that she walks in 3 hours

_____

   **b** the time that she takes to walk 30 km.

_____

**2** Ali has 3.5 hours of maths lessons each week.

Work out the number of hours of maths lessons he has in

   **a** 12 weeks _____

   **b** 38 weeks. _____

**3** Karl has a job with an hourly rate of pay.

His pay is £75.20 for 8 hours.

Work out his pay for

   **a** 1 hour _____      **b** 35 hours. _____

_____       _____

**4** One litre of petrol costs £1.25.

Work out

**a** the cost of 23 litres _____

**b** the number of litres you can buy for £50._____

**5** Work out

**a** the flour you need for 1 scone

_____

**b** the butter you need for 30 scones

_____

**c** the milk you need for 10 scones.

_____

> *Recipe*
>
> Ingredients for 15 scones
>
> Flour     240 g
>
> Butter    45 g
>
> Milk      150 ml

**6** The scale of the plan of a house is 1 : 20.

On the plan the length of a room is 32 cm.

Work out the actual length of the room

**a** in centimetres _____      **b** in metres. _____

**7** 8 km = 5 miles

Work out these missing numbers.

**a** 20 miles = _____ km      **b** 20 km = _____ miles

**8** This is a sales ticket in an airport shop.

**a** Work out how many dollars £1 is worth.

_____

**b** The price of a handbag is £95.

Work out the price in dollars.

_____

> Scarf £25
> or
> $32.50

**9** €1 = £0.90   Work out the missing numbers.

**a** €70 = £ _____      **b** £135 = € _____

       = £ _____            = € _____

## 7.4 Ratios and fractions

**1** In a weather survey, the ratio of sunny days to cloudy days is 3 : 1.

  **a** There were five cloudy days. Work out the number of sunny days.

_____

  **b** Write down the total number of days.

_____

  **c** Work out the fraction of days that are sunny. Write your answer as simply as possible.

_____

**2** Bricks in a wall can be headers ▮ or ▬ stretchers.

  Here is a row of bricks. ▬▬▬▬▬▬▬▬▬▬▬▬

  **a** Write down the ratio of headers to stretchers.

_____

  **b** Write down the **fraction** of the bricks that are headers.

_____

**3** The ratio of tulips to daffodils in a vase of flowers is 1 : 3.
There are 32 flowers altogether.

  **a** Work out the number of each kind.

   Tulips _____       Daffodils _____

  **b** Work out the fraction of these flowers that are tulips.

_____

**4** A box contains only red pens and blue pens. The ratio of the number of red pens to the number of blue pens is 2 : 3.

  Write down the fraction of pencils that are red.

_____

**5** A box contains long and short pencils.
$\frac{3}{5}$ of the pencils are long.

  Write down the ratio of long pencils : short pencils.

_____

**6** There are 20 chickens and 25 ducks.

   **a** Work out the ratio of chickens to ducks. Write the answer as simply as possible.

_____

   **b** Work out the fraction of the birds that are ducks.

_____

**7** Lucy is 120 cm tall. Sam is 40 cm taller than Lucy.

   **a** Work out the ratio of Lucy's height to Sam's height.
     Write the answer as simply as possible.

> Hint: Work out Sam's height.

_____

   **b** Fill in the missing fraction.

     Lucy's height is _____ of Sam's height.

   **c** Fill in the missing fraction.

     Sam's height is _____ of Lucy's height.

**8**

Write your answers as simply as possible.

   **a** Work out the ratio of red squares to white squares.

_____

   **b** Write down the fraction of the squares that are white.

_____

   **c** Write the missing fraction in this sentence.

     The number of white squares is _____ of the number of red squares.

# 7 Problem solving 🖩

**1** A 500 g bag of cereal costs £6.00.

A recommended serving is 25 g.

a Work out the number of servings in one bag.

**Hint:** Divide by 25.

_____

b Work out the cost of one serving.

_____

c Work out the cost of 100 g

_____

John buys a 500 g bag of cereal. He put 200 g in a small box and the rest in a larger box.

d Work out the ratio of the mass in each box. Write your ratio in its simplest form.

**Hint:** Work out the mass in the larger box.

_____

e Write down the fraction of the cereal in the smaller box.

_____

A 380 g bag of the same cereal costs £4.18.

f Work out the cost of 10 g from this bag.

_____

g Work out the cost of 100 g from this bag.

_____

h Which bag is better value, 500 g or 380 g? You must show your working.

_____

_____

**2** A car in France travels 192 km and uses 12 litres of petrol.

a Work out the petrol consumption.
Give your answer in km/l.

**Hint:** Work out 60 ÷ 12.

_____

b Work out the distance the car can travel with 60 litres of petrol.

_____

Petrol costs €1.32 per litre.

c Work out the cost of the petrol for the journey of 192 km. Give your answer in euros.

_____

The exchange rate is £1 = €1.10

d How many euros is £20 worth?

_____

e Work out the cost of the petrol for the journey of 192 km in pounds.

_____

   8 km = 5 miles

f Convert 192 km to miles

_____

g Work out the fuel consumption in miles/l

_____

   4.5 litres = 1 gallon

h Work out the fuel consumption in miles/gallon

_____

**Checklist**

I can

☐ use ratio notation

☐ write a ratio in its simplest form

☐ divide a quantity into two parts in a given ratio

☐ solve simple problems involving proportion

☐ understand the connection between fractions and ratios.

# 8 Percentages

## 8.1 One number as a percentage of another

**1** Complete these equivalent fractions.

Hint: $50 \times ? = 100$.

**a** $\frac{7}{50} = \frac{\square}{100}$ _____

**b** $\frac{7}{25} = \frac{\square}{100}$ _____

**c** $\frac{7}{20} = \frac{\square}{100}$ _____

**d** $\frac{7}{10} = \frac{\square}{100}$ _____

**2** Write these fractions as percentages.

Hint: $\frac{41}{50} = \frac{?}{100}$

**a** $\frac{41}{100} =$ _____

**b** $\frac{41}{50} =$ _____

**c** $\frac{22}{25} =$ _____

**d** $\frac{3}{10} =$ _____

**3** Write these decimals as percentages.

**a** $0.15 =$ _____ %   **b** $0.78 =$ _____ %   **c** $0.03 =$ _____ %

**4** Write the following fractions as percentages. Give your answers to the nearest whole number.

Hint: Find $2 \div 3$ with a calculator.

**a** $\frac{2}{3} =$ _____

**b** $\frac{4}{9} =$ _____

**c** $\frac{7}{9} =$ _____

**d** $\frac{3}{7} =$ _____

**5** There are 40 children. 26 are girls.

Hint: What fraction are girls?

**a** Work out the percentage of girls. _____

**b** Work out the number of boys. _____

**c** Work out the percentage of boys. _____

**6** There 180 children. 117 have brown eyes. 36 have blue eyes.

Work out the percentage with

a brown eyes _____  b blue eyes. _____

**7** There are 360 tickets for a concert. 306 are sold.

a What percentage are sold? _____

b What percentage are not sold? _____

**8** There are three candidates in an election. Here are the results.

| Candidate | Ari | Beth | Carl | Total |
|-----------|-----|------|------|-------|
| **Votes** | 63 | 42 | 27 | 132 |

Work out the percentage for each candidate. Write each answer to the nearest whole number.

Ali _____ Beth _____ Carl _____

**9** During one day 72 trains stop at a station. 57 arrive on time.

a Work out the percentage that are on time. Write each answer to the nearest whole number.

_____

b Write down the percentage that are not on time. Write your answer to the nearest whole number.

_____

# 8.2 Comparisons using percentages 🔢

Key word

percentage

**1** Jen scored the following test marks.

| Subject | Mark | Percentage |
|---------|------|------------|
| **Chinese** | 18 out of 25 | |
| **Engineering** | 132 out of 200 | |
| **Biology** | 52 out of 80 | |

Convert the marks to percentages and complete the table.

**2** a Ros scored 32 out of 40 in a maths test.

Write this as a percentage _____

b Ros scored 42 out of 60 in a science test.

Write this as a percentage _____

c In which subject did she score better, maths or science? _____

**3** In a school year there are 80 boys and 110 girls.

**a** 45% of the boys have a pet.

How many boys have a pet? _____

**b** 44 of the girls have a pet. Show that the percentage who have a pet is smaller for girls than for boys.

_____

_____

**4** Show that 30% is more than $\frac{1}{4}$ but less than $\frac{1}{3}$.

_____

_____

**5** Dan is late for work 6 out of 48 days. Alice is late 8 out of 80 days.

**a** Work out the percentage of days each person is late.

Dan _____                                   Alice _____

_____                                              _____

**b** Who has better attendance? Circle your answer.        Dan        Alice

**6** This table shows the population of two towns.

| Town | 18 and under | Over 18 | Total |
|------|-------------|---------|-------|
| Alfaton | 696 | 1704 | 2400 |
| Betaville | 510 | 990 | 1500 |

**a** Work out the percentage of the Alfaton population that is over 18.

Hint: Use two numbers from each row.

_____

**b** Show that the percentage of the population over 18 is less in Betaville than in Alfaton.

_____

**7** 275 out of 350 students at College A pass an exam.

**a** Work out the percentage that pass. Write your answer to the nearest whole number.

_____

155 out of 185 students at College B pass the same exam.

**b** Show that a greater percentage pass at College B.

_____

# 8.3 Percentage change 🖩

**1** The population of a village was 600. It increases by 25%.

  **a** Work out 25% of 600. _____

  **b** Work out the new population. _____

**2** Last June there was 40 cm of rain. This year there was 60% **less** rain.

  **a** Work out 60% of 40 cm. _____

  **b** Work out the amount of rain this year. _____

  Hint: Do a subtraction.

**3** Jak is paid £15.60 per hour. He gets a pay increase of 5%.

  **a** Work out 5% of £15.60. _____

  **b** Work out his new pay per hour. _____

**4** The cost of a car repair is £320 + VAT.

  The VAT rate is 20%.

  Hint: VAT is a tax.

  **a** Work out 20% of £320. _____

  **b** Work out the total cost including VAT. _____

**5** Last year 5600 people went to a festival. This year there are 35% fewer people.

  **a** Work out 35% of 5600. _____

  **b** Work out how many people went this year. _____

**6** The original price of a coat was £130. In a sale it is reduced by 30%.

  **a** Work out the reduction in pounds. _____

  **b** Work out the sale price. _____

**7** In a sale, prices are reduced by 25%. Complete the following table.

| Item | Original price | Reduction | Sale price |
|---|---|---|---|
| Chair | £240 | £60 | |
| Table | £600 | | |
| Bed | £840 | | |
| Cupboard | | £90 | £270 |

**8** The price of a new car is 50% more than the price of a used car.

The price of the new car is £18 000.

Work out the price of the used car.

> Hint: It is **not** £9000.

## 8 Problem solving  🖩

**1** The following table shows the marks of three students on three tests. It also shows the maximum possible mark on each test.

|  | Test A | Test B | Test C |
|---|---|---|---|
| Ali | 66 | 84 | 102 |
| Emily | 72 | 72 | 80 |
| Fran | 56 | 54 | 87 |
| Maximum mark | **80** | **120** | **150** |

**a** Show that the percentage mark for Ali on test A is 82.5%

> Hint: Use Ali's mark and the maximum for test A.

_____

**b** Work out the percentage marks for Ali on test B and test C.

Test B = _____%        Test C = _____%

**c** Which test has the highest percentage for Fran? Show your working.

_____

_____

**d i** Work out the maximum total mark for the 3 tests.

_____

**ii** Work out Emily's total mark as a percentage of the total possible.

_____

_____

**e** A student with an overall percentage of 70% or more is awarded a distinction. Show that only one of these students has a distinction.

_____

_____

**2** The cost of renting a holiday cottage for a week in August is £625.

When you book it you pay a deposit of £100.

**a** What percentage of the cost is the deposit?

_____

**b** In October there is a 20% reduction based on the August price. Work out the cost in October.

_____

**c** At Christmas there is a 30% increase based on the August price. Work out the cost at Christmas.

_____

A couple book the cottage in August. They estimate they will spend another £800 during the week for petrol, food and entertainment.

**d** Work out the percentage of the total estimated cost of the holiday that is the cottage rent. Give your answer to the nearest whole number.

> Hint: First work out the total estimated cost.

_____

The couple actually spend less than they estimated. The rent of the cottage is 50% of the total cost.

**e** Work out the amount they actually spent on petrol, food and entertainment.

_____

**Checklist**

I can

☐ write one quantity as a percentage of another

☐ use percentages to make comparisons

☐ calculate percentage change

☐ increase or decrease an amount by a percentage.

# 9 Angles and polygons

## 9.1 Points and lines 🖩

The diagrams in this exercise are not drawn accurately. You cannot measure the angles to work out the correct answers.

**Key words**

angle

triangle

corresponding

alternate

opposite

**1** Work out the size of the lettered angles below.

> **Hint:** What is the sum of the angles on a straight line?

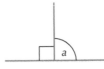

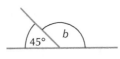

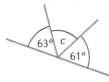

a = _____   b = _____   c = _____

**2** Calculate the size of the lettered angles below.

> **Hint:** What is the sum of the angles around a point?

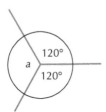

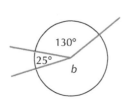

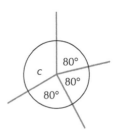

a = _____   b = _____   c = _____

**3**

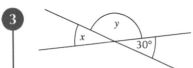

**a** Write down the size of angle x. _____

**b** Work out the size of angle y. _____

**4**

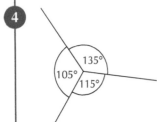

Give a reason why the angles cannot be correct.

_____

**5** Complete the following sentences. Choose the correct word or phrase from this list.

corresponding    alternate    vertically opposite

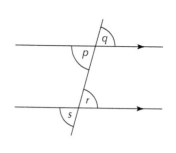

**a** *p* and *q* are _____ angles

**b** *p* and *r* are _____ angles

**c** *p* and *s* are _____ angles

**6** Work out the size of the lettered angles on these parallel lines.

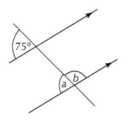

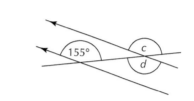

*a* = _____    *b* = _____    *c* = _____    *d* = _____

# 9.2 Triangles 🖩

The diagrams in this exercise are not drawn accurately. You cannot measure the angles to work out the correct answers.

**1** Calculate the size of the lettered angles in these diagrams.

Hint:  What is the sum of the angles of a triangle?

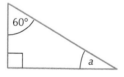

*a* = _____

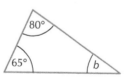

*b* = _____

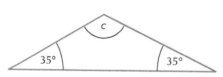

*c* = _____

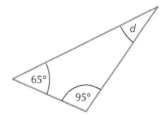

*d* = _____

**2** Work out the value of *a*.

_____

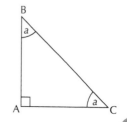

**3** Here is an equilateral triangle inside a rectangle.

Work out the size of the lettered angles.

$x =$ _____ $\qquad$ $y =$ _____

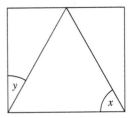

**4**

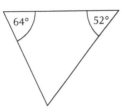

Show that this triangle is isosceles.

_____

_____

**5** PQ = PR = 10 cm. Angle Q = 57°. Work out the sizes
of angle R and angle P.

$R =$ _____ $\qquad$ $P =$ _____

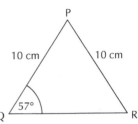

**6** One angle of an isosceles triangle is 130°. Work out the size of the other two angles.
You must show your working.

_____

**7** Work out the size of angle $a$ and angle $b$.

$a =$ _____ $\qquad$ $b =$ _____

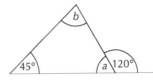

# 9.3 Quadrilaterals and other polygons ▦

**Key words**

quadrilateral

polygon

interior angle

The diagrams in this exercise are not drawn accurately. You cannot
measure the angles to work out the correct answers.

**1** Complete this sentence.

The four angles of a quadrilateral add up to _____.

**2** Work out the size of the lettered angles.

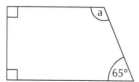

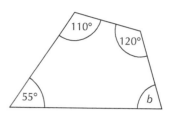

$a =$ _____

$b =$ _____

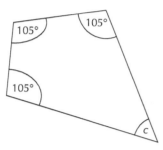

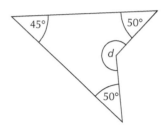

$c =$ _____

$d =$ _____

**3** ABCD is a parallelogram.

**a** Write down the size of angle C. _____

**b** Work out the size of angle D. _____

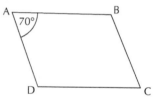

**4**

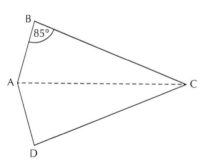

AC is a line of symmetry.

**a** Circle the name of this shape.

     kite               parallelogram             rhombus             trapezium

**b** Write down the size of angle D. _____

**c** Angle A = 150°. Work out the size of angle C. _____

**5** QR and PS are parallel.

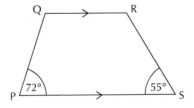

**a** Circle the name of this shape.

   kite         parallelogram

   rhombus     trapezium

**b** Angle P = 72°. Work out the size of angle Q. _____

**c** Angle S = 55°. Work out the size of angle R. _____

**6** A rectangle is removed from a square to make this shape.

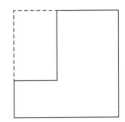

**a** Circle the name of the shape.

   pentagon      hexagon      octagon      quadrilateral

**b** Work out the sum of the interior angles of the shape.

> **Hint:** One angle is different from all the others.

_____

**7** This shape is made from a rectangle and an equilateral triangle.

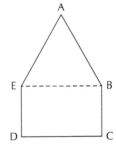

**a** Circle the name of the shape.

   pentagon      hexagon      octagon      quadrilateral

**b** Work out the size of angle E. _____

**c** Work out the sum of all of the interior angles. _____

# 9.4 Shapes on coordinate axes

**Key words**

symmetry

vertex

**1**

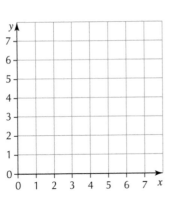

ABCD is a rectangle.

**a** Plot the points A(2, 2) , B(6, 2) and C(6, 4).

**b** Draw the rectangle.

**c** Write down the coordinates of D. _____

**d** The rectangle has two lines of symmetry. Write down their equations.

_____

**2** PQRS is a parallelogram.

**a** Plot P(2, 3), Q(2, 6) and R(5, 3).

**b** Draw PQ and QR.

**c** Complete the parallelogram.

**d** Write down the coordinates of S. _____

**e** Work out the size of the interior angles of the parallelogram.

_____

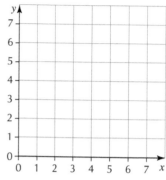

**3** **a** Draw the line $y = 4$.

**b** Plot A(1, 4), B(3, 6) and C(7, 4).
ABCD is a kite and $y = 4$ is the
line of symmetry of the kite.

**c** Draw the kite.

**d** Write down the coordinates of D. _____

**e** Work out the size of angle BAD.

_____

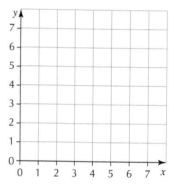

**4** **a** Plot A(4, 0) and C(4, 6).
A and C are opposite vertices of the square ABCD.

**b** Draw the square.

Hint: Start with the centre of the square.

**c** Write down the coordinates of B and D.

_____

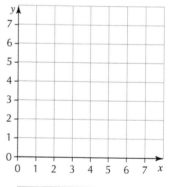

**5** (4, 3) is the centre of a square. One vertex is at (3, 0).

**a** Plot (4, 3) and (3, 0).

**b** Draw the square.

Hint: Start with the vertex opposite (3, 0).

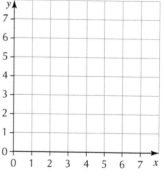

**6**

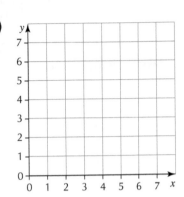

**a** Draw the line $y = x$.

Hint: It is a diagonal line.

ABCDEF is a hexagon.

**b** Plot A(2, 2), B(1, 4), C(3, 6) and D(5, 5).

$y = x$ is a line of symmetry of the hexagon.

**c** Draw the hexagon.

Hint: Draw the reflection of the points you know.

**d** There is another line of symmetry. Write down its equation. _____

# 9 Problem solving

**1** This question is about the following two triangles.

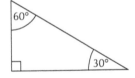 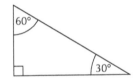 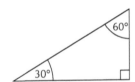

You can also reflect them like this.

**a** Put the two triangles together to make a rectangle.

Hint: Join two edges together.

**b** Put the two triangles together to make an equilateral triangle.

**c** Put the two triangles together to make an isosceles triangle.

**d** Write down the sizes of the interior angles of the isosceles triangle.

_____

**e** Put the two triangles together to make a parallelogram.

**f** Write down the sizes of the interior angles of the parallelogram. _____

**g** Put the two triangles together to make a kite.

**h** Write down the sizes of the interior angles of the kite. _____

**2**

Hint: Look at the angles at P.

This pattern is made from regular hexagons.

**a** How many hexagons meet at point P?

_____

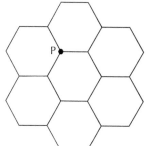

**b** Explain why the angles of a regular hexagon must be 120°.

_____

This pattern also contains regular hexagons.

**c** How many shapes meet at Q?

_____

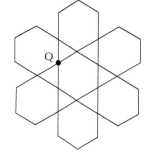

**d** Explain why the triangles must be equilateral.

_____

_____

This pattern also contains regular hexagons.

**e** Write down the name of the other shape in the pattern.

_____

**f** Work out the angles of the other shape.

_____

Here is a regular octagon divided into smaller shapes.

**g** Work out the size of each angle of the regular octagon. Explain how you know.

_____

_____

**Checklist**

I can

☐ work out angles on a straight line

☐ work out angles round a point

☐ work out angles in parallel lines

☐ work out angles in a triangle

☐ work out angles in a quadrilateral

☐ name different types of polygons.

# 10 Perimeter, area and volume

## 10.1 Perimeter and area

**1** Work out the perimeter of each shape.

> **Hint:** Perimeter = length of the boundary.

**a**

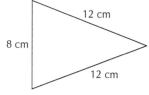

12 cm

8 cm

12 cm

_____ cm

**b**

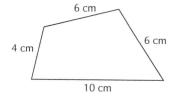

6 cm

4 cm

6 cm

10 cm

_____ cm

**2** Work out the perimeter of each rectangle.

**a**

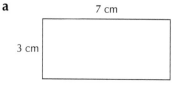

7 cm

3 cm

_____ cm

**b**

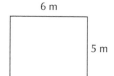

6 m

5 m

_____ m

**3** Work out the area of each rectangle in question 2.

**a** _____ cm²

**b** _____ m²

**4** The perimeter of a square is 20 cm.

**a** Work out the length of each side.

_____

**b** Work out the area of the square.

_____

**5**

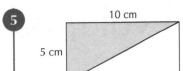

This is a rectangle.

**a** Work out the area of the rectangle.

_____

**b** Work out the area of the triangle. Hint: What fraction of the rectangle is the triangle?

_____

**6** Work out the area of each triangle.

**a**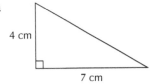

**b**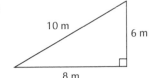

_____ cm² _____ m²

**7** Work out the perimeter of each of these parallelograms.

**a**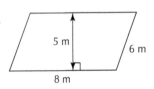

**b**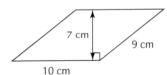

_____ m _____ cm

**8** Work out the **area** of each of the parallelograms in question 7. Hint: Area = base × height.

**a** _____ m² **b** _____ cm²

**9** This shape is made from two rectangles.

All lengths are in centimetres.

**a** Work out the perimeter of the shape.

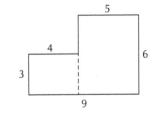

_____

**b** Work out the area of the shape. Hint: Work out the area of each rectangle separately.

_____

**10** Here is an L-shape.

All lengths are in centimetres.

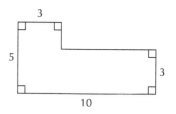

**a** Work out the perimeter of the shape.

_____

**b** Work out the area of the shape.

_____

## 10.2 Circles

**Key words**

circumference

area

**1**

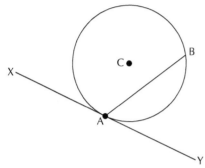

C is the centre of the circle.

**a** Circle the correct name for the straight line AB.

diameter          chord          sector          tangent

**b** Circle the correct name for the straight line XAY.

diameter          chord          sector          tangent

**2**

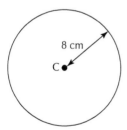

C is the centre of the circle.

**a** Work out the diameter of the circle.

_____

**b** Calculate the circumference. Give your answer to the nearest centimetre.

_____

**3** C is the centre of the circle.

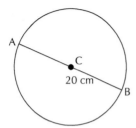

AB = 20 cm.

Calculate the circumference. Give your answer to the nearest centimetre.

_____

**4** The circumference of a circle is 1 m.

Work out the diameter to the nearest centimetre. Circle the correct answer.

8 cm          1 cm          32 cm          50 cm

**5** Calculate the circumference of each of these circles. Give your answers to the nearest centimetre.

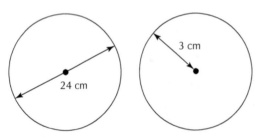

**a** _____ cm   **b** _____ cm

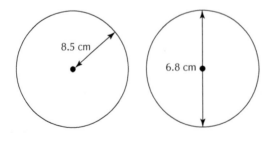

**c** _____ cm   **d** _____ cm

**6**

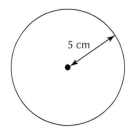

5 cm

The radius of this circle is 5 cm.

Calculate the area of the circle.

Give your answer to the nearest whole number.

Show your working.

_____

**7** Calculate the area of each of the circles in question **5**.

a _____ cm²          b _____ cm²

c _____ cm²          d _____ cm²

**8**

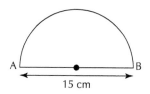

A                    B

15 cm

This shape is half a circle.

**a** Calculate the length of the arc AB, to the nearest centimetre.

Hint:   An arc is part of the circumference.

_____

**b** Calculate the area of the sector, to the nearest cm².

_____

# 10.3 Solids

**1**

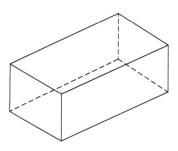

This is a cuboid. Write down the number of

**a** faces _____    **b** edges _____    **c** vertices. _____

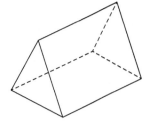

**2** This is a triangular prism. Write down the number of

**a** faces _____ **b** edges _____ **c** vertices. _____

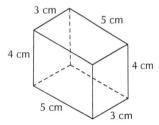

**3** This is a cuboid. Work out the area of

**a** the largest face _____ **b** the smallest face. _____

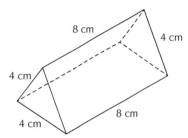

**4** This is a triangular prism.
Work out the area of one of the rectangular faces.

_____

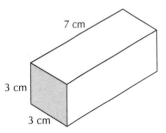

**5** The coloured face of this cuboid is a square. Work out

**a** the area of the square _____ cm² **b** the volume of the cuboid. _____ cm³

**6** Calculate the volume of each of these cuboids.

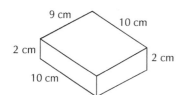

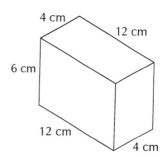

**a** _____ cm³        **b** _____ cm³

**7**

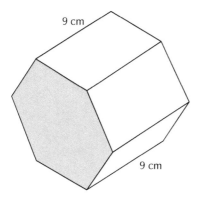

This is a hexagonal prism.

**a** Write down the number of

   **i** faces _____   **ii** edges _____   **iii** vertices. _____

**b** The area of the coloured end is 30 cm². The length is 9 cm. Calculate the volume.

_____

_____

**8** Each edge of a cube is 8 cm. Calculate

**a** the area of one face _____ cm²

**b** the volume. _____ cm³

# 10 Problem solving

**1** This is a rectangular tile T.

**a** Work out the perimeter of the tile.

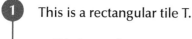

**b** Work out the area of the tile.

_____

Two tiles can be put together to make a larger rectangle as shown.

**c** Work out the perimeter of the larger rectangle.

_____

| T |
|---|
| T |

**d** Work out the area of the larger rectangle.

_____

**e** Show how two of the tiles can be put together to make a different rectangle.

Hint: Your drawing does not need to use the accurate measurements.

**f** For the rectangle in part **e** work out

**i** the perimeter _____        **ii** the area. _____

**g** Draw a different rectangular tile that has the same **area** as tile T. Write the lengths of the sides on your diagram.

**h** Work out the perimeter of the tile in part **g**.

_____

**i** A square tile has the same perimeter as tile T. Work out the area of the square tile.

_____

_____

**2** This is a cube. Each edge has length 3 cm.

**a** Work out the area of one face.

_____

3 cm    3 cm
3 cm

**b** Work out the total area of all the faces.

Hint: There are 6 faces.

_____

**c** Work out the volume of the cube.

_____

**d** Work out the total length of all the edges.

Hint: There are 12 edges.

_____

Two cubes are put together to make a cuboid.

**e** Work out the volume of the cuboid.

_____

**f** Work out the area of one rectangular face of the cuboid.

_____

**g** Work out the total area of all the faces of the cuboid.

_____

**h** Work out the total length of all the edges.

_____

10 cubes

10 cubes are put in a line to make a long cuboid.

**i** Work out the total length of all the edges.

_____

**Checklist**

I can

☐ work out the perimeter of a shape

☐ work out the area of a rectangle, a triangle and a parallelogram

☐ work out the circumference and area of a circle

☐ find the number of vertices, faces and edges on a simple solid

☐ work out the volume of a cuboid.

# 11 Transformations

## 11.1 Reflections

1  Draw the reflection of each shape in the mirror line.

a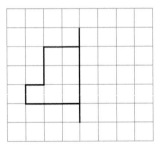

b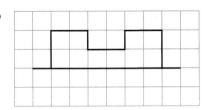

2  Draw the reflection of each shape in the mirror line.

a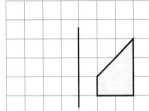

b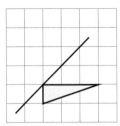

3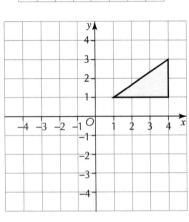

a  Draw the reflection of the triangle in the *x*-axis. Label it A.

b  Draw the reflection of the triangle in the *y*-axis. Label it B.

4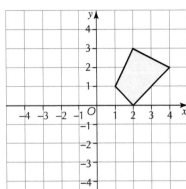

a  Draw the reflection of the quadrilateral in the *x*-axis. Label it A.

b  Draw the reflection of the quadrilateral in the *y*-axis. Label it B.

**5**

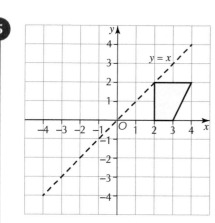

a Draw the reflection of the shape in the x-axis.
Label it A.

b Draw the reflection of the shape in the y-axis.
Label it B.

c Draw the reflection of the shape in line y = x.
Label it C.

Hint: The line y = x is drawn for you.

**6**

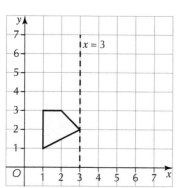

a Draw the reflection of the shape in the line x = 3.
Label it A.

b Draw the line y = 3.

c Draw the reflection of the shape in the y = 3.
Label it B.

**7**

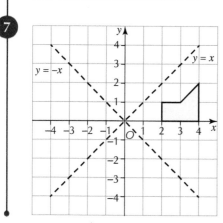

a Draw the reflection of the shape in line y = x.
Label it A.

b Draw the reflection of the shape in line y = − x.
Label it B.

Hint: The line is on the grid.

## 11.2 Translations and rotations

**1**

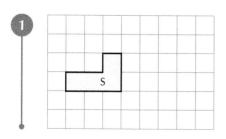

**Key words**

translation

vector

image

rotation

**a** $\begin{pmatrix} 4 \\ 2 \end{pmatrix}$ means 4 squares to the right and 2 squares up

Translate S by the vector $\begin{pmatrix} 4 \\ 2 \end{pmatrix}$. Label the image A.

**b** $\begin{pmatrix} 4 \\ -2 \end{pmatrix}$ means 4 squares to the right and 2 squares down

Translate S by the vector $\begin{pmatrix} 4 \\ -2 \end{pmatrix}$. Label the image B.

**c** Write down the vector for the translation from B to A.

_____

**2**

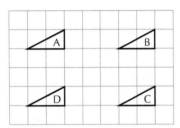

Write down the vectors for these translations.

**a** D to B _____ **b** B to A _____ **c** A to C _____

**3** Rotate triangle ABC

   **a** 180° about A

   **b** 180° about B

   **c** 180° about C.

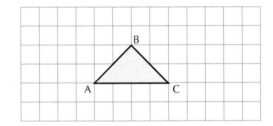

**4**   **a** Rotate the triangle 90° clockwise about X.

    **b** Rotate the triangle 90° anticlockwise about Z.

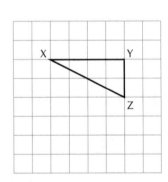

**5**

**a** Translate P by the vector $\begin{pmatrix} -2 \\ 3 \end{pmatrix}$.

**b** Rotate P 180° about (0, 0).

**c** Rotate P 90° clockwise about (0, 0).

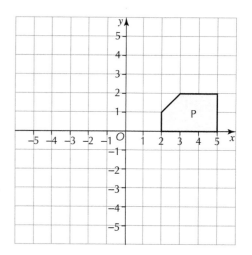

**6**

**a** Rotate R 90° clockwise about (0, 0). Label the image S.

**b** Rotate S 90° clockwise about (0, 0). Label the image T.

**c** Translate R by the vector $\begin{pmatrix} -4 \\ -2 \end{pmatrix}$. Label the image U.

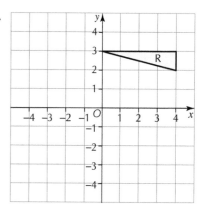

**7**

**a** Rotate T 180° about (5, 5). Label the image U.

Hint: (5, 5) is one of the vertices of the triangle.

**b** Rotate T 180° about (5, 3). Label the image V.

**c** Write down the vector for the translation from U to V.

**d** Translate T by the vector $\begin{pmatrix} -1 \\ -3 \end{pmatrix}$. Label the image W.

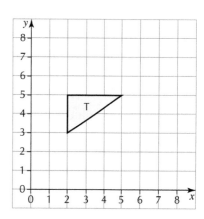

# 11.3 Enlargements

**Key words**

enlargement

scale factor

centre

**1**

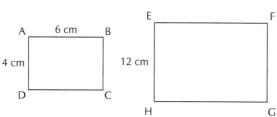

Not drawn accurately

ABCD is a rectangle. EFGH is an enlargement of rectangle ABCD.

**a** Work out the scale factor of the enlargement. _____

**b** Work out the length EF. _____

**2**

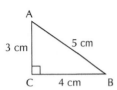

Not drawn accurately

Triangle ABC is enlarged to triangle DEF.

**a** Work out the scale factor of the enlargement. _____

**b** Work out the perimeter of triangle DEF. _____

**3** The perimeter of this shape is 30 cm.

It is enlarged with a scale factor of 6.

Work out the perimeter of the enlargement. _____

**4** EFGH is an enlargement of ABCD.

The scale factor is 2.

O is the centre of the enlargement.

EF has been drawn. Complete the enlargement of EFGH.

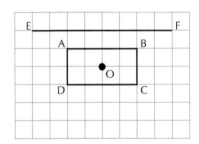

**5** One side of an enlargement of this triangle has been drawn.

The centre is X. The scale factor is 3.

Complete the enlargement.

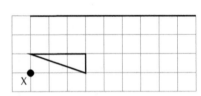

**6** Draw an enlargement of this square, centre (0, 0), scale factor 2.

Hint: Double the coordinates of the vertices.

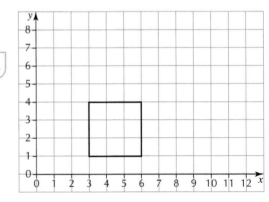

**7** Draw an enlargement of this rhombus, centre (0, 0), scale factor 3.

> Hint: Multiply the coordinates of the vertices by 3.

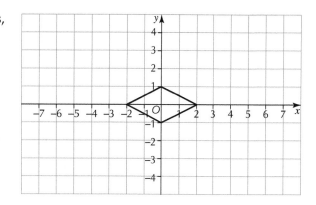

**8** Draw an enlargement of this triangle, centre (0, 0), scale factor 2.

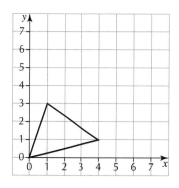

# 11 Problem solving

Key word

image

**1** **a** Reflect triangle A in the x-axis. Label the image B.

> Hint: The image is the new triangle.

**b** Reflect triangle A in the y-axis. Label the image C.

**c** Rotate triangle A 180° about (0, 0). Label the image D.

**d** Reflect triangle E in the y-axis. Label the image F.

**e** Rotate triangle E 180° about (0, 0). Label the image G.

**f** Reflect triangle E in the x-axis. Label the image H.

**g** Describe the transformation that maps triangle E to triangle A.

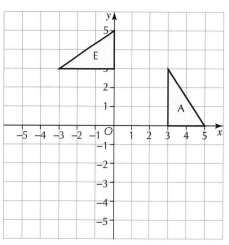

_____

**h** Translate triangle E by the vector $\begin{pmatrix} 0 \\ -6 \end{pmatrix}$. Label the image J.

**i** Describe the transformation that maps triangle J to triangle H.

_____

**2**

**a** Draw the enlargement of triangle A, centre X, scale factor 2.

Two vertices of the enlargement have been drawn on the grid. Label the enlargement B.

**b** Draw the enlargement of triangle B, centre X, scale factor 2. Label it C.

The longest side of A has length 6 cm.

**c** Work out the length of the longest side of triangle B.

_____

**d** Work out the length of the longest side of triangle C.

_____

**e** Triangle C is an enlargement of triangle A.

Work out the scale factor of the enlargement. _____

**f** On your drawing, show that triangle B can be divided into triangles the same size and shape as triangle A.

The area of triangle A is 9 cm².

**g** Work out the area of triangle B. _____

**h** Work out the area of triangle C. _____

---

**Checklist**

I can

☐ draw the reflection of a shape

☐ rotate a shape 90° or 180° about a given centre

☐ translate a shape by a given vector

☐ enlarge a shape with a given scale factor

☐ describe different types of transformations.

# 12 Probability

## 12.1 Relative frequency

**Key words**

frequency

relative frequency

outcome

**1** A drawing pin is dropped 20 times. It lands point up or point down.

| Point | Up | Down | Total |
|---|---|---|---|
| Frequency | 13 | 7 | 20 |

Write down the relative frequency of

**a** point up _____

**b** point down. _____

> Hint: Relative frequency = $\dfrac{\text{number of favourable events}}{\text{total number of events}}$.

**2** Two coins are spun together 50 times. Here are the results.

| Outcome | 2 heads | 2 tails | 1 head, 1 tail |
|---|---|---|---|
| Frequency | 9 | 14 | 27 |

Work out the relative frequency of

**a** 2 heads _____

**b** the same on both coins. _____

**3** Vehicles travelling along a road in one hour are recorded.

| Vehicle | Car | Van | Lorry |
|---|---|---|---|
| Frequency | 154 | 32 | 14 |

**a** Work out the total number of vehicles.

_____

**b** Work out each of these relative frequencies. Write your answers as percentages.

**i** car _____

**ii** lorry _____

**iii** not a lorry _____

**4** Ahmed throws a dice 20 times. Here are his scores.

2   3   1   5   5   4   5   1   3   2   5   3   6   2   4   6   5   2   3   3

**a** Complete the table of his scores.

| Score | 1 | 2 | 3 | 4 | 5 | 6 | Total |
|---|---|---|---|---|---|---|---|
| Frequency | 2 | 4 | 5 | | | | 20 |

**b** Work out the relative frequencies of each of these results. Give your answers as decimals.

**i** 2 _____

**ii** 6 _____

**iii** an odd number _____

**5** A computer simulation throws two dice 1000 times.

The frequency of a total of 8 or more is 427.

**a** Work out the relative frequency of a total of less than 8.

_____

The relative frequency of double 6 is 0.04.

**b** Work out the frequency of double 6.

> Hint: Frequency = relative frequency × total number of times the two dice are thrown.

_____

**6** A survey of 200 people records age and gender as shown in the following table.

|  | Under 18 | 18–60 | Over 60 | Total |
|---|---|---|---|---|
| **Male** | 12 | 42 | 26 | 80 |
| **Female** | 20 | 75 | 25 | 120 |
| **Total** | 32 | 117 | 51 | 200 |

**a** Work out the relative frequency of females.

_____

**b** For females only, work out the relative frequency of under 18s.

_____

**c** For under 18s only, work out the relative frequency of females.

_____

# 12.2 Calculating probabilities

**Key words**

random

fair

equally likely

**1** Here are eight number cards. One is chosen at random.

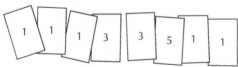

Put a letter on the probability scale below for each of the outcomes.

**A:** the card is 1    **B:** the card is less than 5    **C:** the card is an even number

0                                                                    1

**2** Carla throws an ordinary fair dice. Work out the probability she throws

**a** 2 _____    **b** more than 2 _____    **c** an odd number. _____

**3** 15 men, 30 women and 5 children each buy one raffle ticket. There is one prize.

Work out the probability that the winner is

**a** a woman _____     **b** not a woman _____     **c** not a child. _____

**4** There are 10 balls in a box, 4 red, 3 blue, 2 green and 1 yellow.

A ball is taken out at random.

**a** Work out the probability that the ball is

    **i** red _____     **ii** blue or green _____     **iii** not yellow. _____

The yellow ball is taken from the box.

A second ball is taken at random from the box.

**b** Work out the probability that the second ball is

    **i** red _____     **ii** blue or green _____     **iii** not yellow. _____

**5** Sam catches the bus each day for 60 days. He records how late it is.

| | On time | Up to 5 minutes late | 5 to 10 minutes late | Over 10 minutes late |
|---|---|---|---|---|
| **Frequency** | 40 | 10 | 6 | 4 |

Use the table to find the probability that the bus will be

**a** on time _____     **b** late _____     **c** at least 5 minutes late. _____

**6** The probability that Carrie beats Darren at badminton is 0.3.

They play 20 games. How many can Carrie expect to win?

> Hint: Expected number of wins = probability of win × number of games played.

**7** A spinner has three sections of different sizes and colours.

Here are the results of 25 spins.

| **Colour** | Green | Blue | Red | Total |
|---|---|---|---|---|
| **Frequency** | 5 | 12 | 8 | 25 |

**a** Use the results to write the probability of each colour.

    **i** green _____     **ii** blue _____     **iii** red _____

**b** The spinner is spun 120 times. Work out the expected number of greens.

> Hint: Expected number of greens = probability of green × total number of spins.

## 12.3 Possibility spaces

**Key words**

possibility

outcome

**1** In a café there are three choices of main courses:
pie (A), risotto (B), pasta (C).

There are two choices of puddings: fruit (X), yogurt (Y).

Jon chooses a main and a pudding. One possible choice is AX.

**a** List the other possible choices in the same way.

_____

A third pudding is added to the list: ice cream (Z). There is no risotto left.

**b** List all the possible choices now.

_____

**2** An ordinary fair dice is thrown and a fair coin is spun.

**a** Complete this table of outcomes.

**Dice**

|  |  | 1 | 2 | 3 | 4 | 5 | 6 |
|---|---|---|---|---|---|---|---|
| **Coin** | **Head (H)** | 1H | | | | | |
|  | **Tail (T)** | | 2T | | | | |

**b** Write down the number of different outcomes.

_____

**c** Work out the probability that Sonia throws

**i** 2 and a tail _____      **ii** an even number and a head. _____

**3** This spinner has three equal sectors red (R), yellow (Y) and blue (B).

**a** The arrow is spun twice. Complete this table of possible outcomes.

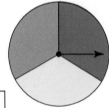

| | | **First spin** | | |
|---|---|---|---|---|
| | | Red (R) | Yellow (Y) | Blue (B) |
| **Second spin** | **Red (R)** | | | |
| | **Yellow (Y)** | | | BY |
| | **Blue (B)** | RB | | |

**b** Write down the number of possible outcomes.

_____

**c** Work out the probability of

**i** 2 reds _____

**ii** the same colour on each spin _____

**iii** different colours on each spin. _____

**4** An ordinary fair dice is thrown twice. Here is a possibility space diagram.

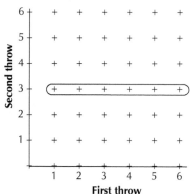

**a** Write down the number of outcomes.

_____

The loop shows the event '3 on the second throw'.

**b** Work out the probability of three on the second throw.

_____

**c** Work out the probability of

   **i** 6 on both throws _____

   **ii** the same number on both throws _____

   **iii** an even number on both throws _____

   **iv** at least one 6. _____

**5** Two ordinary fair dice are thrown. The scores are added.

**a** Complete this table of possible totals.

**First dice**

| | | 1 | 2 | 3 | 4 | 5 | 6 |
|---|---|---|---|---|---|---|---|
| | **6** | | | | | 11 | |
| | **5** | | | | | | |
| **Second** | **4** | | | | 8 | | |
| **dice** | **3** | | | | | | |
| | **2** | 3 | | | | | |
| | **1** | | | | | | |

**b** Write down the most likely total.

_____

**c** Write down the least likely totals.　　　　**Hint:** There are two possible answers.

_____

**d** Work out the probability of a total of

   **i** 12 _____     **ii** 3 _____

   **iii** 8 _____     **iv** 10 or more. _____

**1** A spinner has 5 equal sectors.

The arrow is spun 20 times.

Here are the results.

| Colour | Green | Yellow | Red | Total |
|---|---|---|---|---|
| Frequency | 6 | 3 | 11 | 20 |

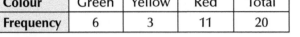

**a** Work out the relative frequency of each colour as a decimal.

> Hint: Write it as a fraction first.

   **i** green _____    **ii** yellow _____    **iii** red _____

The arrow is now spun 30 more times.

Here are the results.

| Colour | Green | Yellow | Red | Total |
|---|---|---|---|---|
| Frequency | 7 | 6 | 17 | 30 |

**b** Combine the results of all 50 spins into a single table.

| Colour | Green | Yellow | Red | Total |
|---|---|---|---|---|
| Frequency | | | | 50 |

**c** Work out the following relative frequencies for the combined results, writing your answers as decimals.

   **i** green _____    **ii** yellow _____    **iii** red _____

**d** Assume the spinner is fair. Write the probability of each colour as a decimal.

   **i** green _____    **ii** yellow _____    **iii** red _____

**e** Tom says "The spinner is not fair because the relative frequencies and the probabilities are different".

Do you agree?       Yes       No

Give a reason for your answer.

_____

**2** Two fair coins are spun.

This outcome is HT.

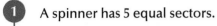

**a** Complete this table to show all the possible outcomes.

|  |  | Coin 1 | |
|---|---|---|---|
|  |  | Head (H) | Tail (T) |
| Coin 2 | Head (H) | | |
|  | Tail (T) | HT | |

**b** Work out the probability of

  **i** 2 heads _____

  **ii** 1 head and 1 tail. _____

  > Hint: The answer to part **ii** is not $\frac{1}{4}$

Now **three** fair coins are spun.

This outcome is HTT.

**c** Complete this table to show all the possible outcomes.

<table>
<tr><th colspan="2"></th><th colspan="4">Coins 1 and 2</th></tr>
<tr><th colspan="2"></th><th>HH</th><th>HT</th><th>TH</th><th>TT</th></tr>
<tr><td rowspan="2">Coin 3</td><td>Head (H)</td><td></td><td></td><td></td><td>TTH</td></tr>
<tr><td>Tail (T)</td><td></td><td>HTT</td><td></td><td></td></tr>
</table>

**d** Work out the probability of

  **i** 3 heads _____

  **ii** 2 heads and 1 tail _____

  **iii** 1 heads and 2 tails _____

Now four coins are spun.

**e** Complete this table to show all the possible outcomes.

<table>
<tr><th colspan="2"></th><th colspan="8">Coins 1, 2 and 3</th></tr>
<tr><th colspan="2"></th><th>HHH</th><th></th><th></th><th></th><th></th><th></th><th></th><th></th></tr>
<tr><td rowspan="2">Coin 4</td><td>Head (H)</td><td></td><td></td><td></td><td></td><td></td><td></td><td></td><td></td></tr>
<tr><td>Tail (T)</td><td>HHHT</td><td></td><td></td><td></td><td></td><td></td><td></td><td></td></tr>
</table>

**f** Work out the probability of

  **i** 4 heads _____

  **ii** 3 heads and 1 tail _____

  **iii** 2 heads and 2 tails. _____

**Checklist**

I can

- [ ] list outcomes systematically
- [ ] work out relative frequencies
- [ ] work out probabilities based on equally likely outcomes
- [ ] use tables and probability spaces for combined outcomes.

# 13 Statistics

## 13.1 Tables, charts and diagrams  🖩

**1** Monty is comparing car insurance prices.

There are two types of car insurance: comprehensive and third party.

Here are the prices for four companies.

| Company | Comprehensive | Third Party |
|---------|---------------|-------------|
| A | £280 | £225 |
| B | £324 | £247 |
| C | £295 | £219 |
| D | £307 | £253 |

**a** Which company is most expensive for comprehensive insurance?

_____

**b** Which company is cheapest for third party insurance?

_____

**c** Work out the difference between the two prices for company D.

_____

**2** This pictogram shows the number of calls to a telephone helpline.

**Telephone calls to helpline**

Monday      ◯ ◯ ◯ ◯

Tuesday     ◯ ◯ ◖

Wednesday

There were 40 calls on Monday.

**a** Work out the number of calls on Tuesday.

_____

**b** There were 15 calls on Wednesday. Show this on the pictogram.

_____

**3** This bar chart shows the populations of six countries in Europe.

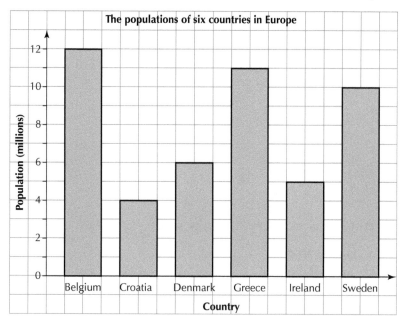

**a** Write down the country with the largest population.

_____

**b** Work out the difference between the populations of Greece and Ireland.

_____

**c** Work out the total population of the six countries.

_____

**4** This graph shows the average temperature in Paris each month.

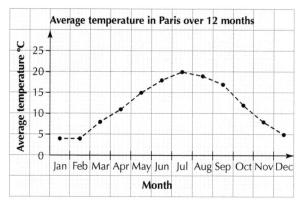

**a** Name the coldest months.

_____

**b** Name the month with an average temperature of 15°C.

_____

**c** Work out the number of months with an average temperature below 15°C.

_____

**5** New Zealand won 18 medals at the 2016 Olympics.
How many medals were

**New Zealand medals at the 2016 Olympics**

**a** silver _____

**b** gold _____

**c** bronze. _____

# 13.2 Calculating statistics 🖩

**Key words**

mode

median

range

mean

frequency table

**1** A group of students were asked their favourite sport.

**Favourite Sports**

**Favourite Sports**

Write down the modal favourite sport for

Hint: The mode is the most common.

**a** girls _____        **b** boys. _____

**2** Here are 25 ages in years.

| 11 | 12 | 12 | 13 | 13 | 13 | 13 | 14 | 14 | 15 | 15 | 15 | 15 | 15 | 16 |

| 16 | 17 | 17 | 17 | 17 | 17 | 17 | 18 | 20 | 20 |

Work out

**a** the modal age _____     **b** the median _____     **c** the range. _____

**3** Here are the hourly rates of pay for 9 people.

£8.50     £8.50     £8.50     £9.30     £9.30     £10.15     £10.80     £12.40     £14.80

**a** Work out

**i** the median _____        **ii** the range. _____

Two more people are added to the list. They both have hourly rates of £12.25.

**b** For all 11 people work out

**i** the median _____        **ii** the range. _____

**4** Here are the number of children in 40 families.

0  0  0  0  0  0  1  1  1  1  1  1  1  1  1  1  1  1  2  2
2  2  2  2  2  2  2  3  3  3  3  3  3  3  4  4  5  5  5  7

**a** Complete the following frequency table.

| Number of children | 0 | 1 | 2 | 3 | 4 | 5 | 6 | 7 |
|---|---|---|---|---|---|---|---|---|
| Frequency | 6 | 12 | 9 | | | | | |

**b** Work out

**i** the mode _____    **ii** the median _____    **iii** the range. _____

**5** Here are the times five people wait in a queue.

3 minutes    7 minutes    2 minutes    10 minutes    2 minute

**a** Work out the range.

_____

**b** Work out the mean waiting time in minutes.

_____

**6** Here are the marks for six students for two exam papers.

| Student | Ali | Beth | Carl | Dima | Ethan | Finn |
|---|---|---|---|---|---|---|
| Paper 1 | 28 | 29 | 41 | 18 | 30 | 29 |
| Paper 2 | 24 | 30 | 33 | 28 | 41 | 36 |

**a** Work out the range for Paper 1.

_____

**b** Work out the mean mark for Carl.

_____

**c** Work out the mean mark for Paper 2.

_____

**7** Here are the number of hours of sunshine each day in one week in Seatown.

| Day | Sun | Mon | Tue | Wed | Thu | Fri | Sat |
|---|---|---|---|---|---|---|---|
| Hours | 2 | 2 | 4 | 7 | 8 | 8 | 4 |

**a** Work out

**i** the range _____    **ii** the median _____    **iii** the mean. _____

In Lakeside the mean amount of sunshine in the same week was 3 hours a day.

**b** Work out the total number of hours sunshine in Lakeside.

_____

# 13 Problem solving 🖩

**1** The following bar chart shows the energy content of 5 types of baked goods.

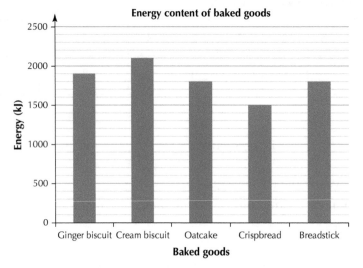

**Hint:** Energy is measured in kJ.

**a** Write down the energy content of ginger biscuits _____ kJ

**b** Write down the type of baked good with the least energy content.

_____

**c** Work out the difference in energy content between oatcakes and crispbread.

_____ kJ

This chart shows the protein and fat content of 4 types of baked goods.

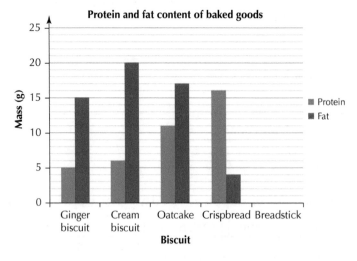

**d** Breadsticks have 14 g of protein and 9 g of fat. Add these to the chart.

**e** Work out the median protein content.

_____ g

**f** Work out the range of the fat content.

_____ g

This pie chart shows the carbohydrate content of the baked goods.

**Carbohydrate content of the baked goods**

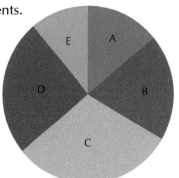

**g** Write the correct letter (A, B, C, D or E) after these statements.

**i** The largest carbohydrate content is ginger biscuits,

letter _____

**ii** The smallest carbohydrate content is oatcake,

letter _____

**iii** The median carbohydrate content is crispbread,

letter _____

**2** Five people each carried out three tasks. Here are their times in minutes.

|  | Ian | Joe | Kim | Lisa | Mika |
|---|---|---|---|---|---|
| **Task 1** | 12 | 18 | 20 | 14 | 11 |
| **Task 2** | 16 | 14 | 13 | 22 | 17 |
| **Task 3** | 21 | 23 | 18 | 15 | 18 |

**a** Work out the median time for task 1.

_____

**b** Work out the range for task 2.

_____

**c** Work out the mean time for task 3.

_____

**d** Work out the mean time for the three tasks for Kim.

_____


40 people carry out task 4. Here are their times in minutes.

8   8   8   8   8   8   9   9   9   9   9   10   10   11   11   11   11   11   11   12
12   12   12   12   13   13   13   13   13   14   14   14   14   14   14   14   14   14
14   14

**e** Complete the following frequency table for Task 4.

| Time (minutes) | 8 | 9 | 10 | 11 | 12 | 13 | 14 |
|---|---|---|---|---|---|---|---|
| Frequency | 6 | 5 |  | 6 |  |  | 11 |

**f** For Task 4 work out

**i** the range _____    **ii** the modal time _____    **iii** the median time. _____

**g** All the times for Task 4 add up to 458 minutes. Work out the mean time for Task 4.

_____

100 people do Task 5. The total of all their times is 875 minutes.

**h** On average, which task was completed more quickly?

Circle your answer.     Task 4          Task 5

Give a reason for your answer.

_____

Checklist

I can

☐ interpret and draw charts and diagrams

☐ interpret and construct frequency tables

☐ work out the median, mode and mean for a set of data

☐ work out the range of a set of data.

# Glossary

**24-hour clock:** Measuring the time based on all 24 hours of the day.

**Alternate (angles):** Angles that lie on either side of a line that cuts a pair of parallel lines; the line forms two pairs of alternate angles and the angles in each pair are equal.

**Angle:** The space between two straight lines that diverge from a common point or between two planes that extend from a common line.

**Area:** The extent of a two-dimensional surface enclosed within a specified boundary or geometric figure.

**Axis:** One of the two lines on a graph on which the scales of measurement are marked.

**Bar chart:** A diagram which is a series of bars or blocks of the same width to show frequencies.

**Bracket:** A pair of marks that are placed around a series of symbols in a mathematical expression to indicate that those symbols function as one item within the expression.

**Centre:** The midpoint of any line or figure, especially the point within a circle or sphere that is equidistant from any point on the circumference or surface; a point, area, or part that is approximately in the middle of a larger area or volume; the point, axis or pivot about which a body rotates.

**Circumference:** The perimeter of a circle; every point on the circumference is the same distance from the centre, and this distance is the radius.

**Common factor:** A whole number that divides exactly into two or more numbers; 2 is a common factor of 6, 8 and 10.

**Common multiple:** An integer or polynomial that is a multiple of each integer or polynomial in a group.

**Coordinates:** The two sets of numbers or letters on a map or graph that you need in order the find that point.

**Corresponding (angles):** Angles that lie on the same side of a pair of parallel lines cut by a line; the line forms four pairs of corresponding angles, and the angles in each pair are equal.

**Cube:** A solid having six plane square faces in which the angle between two adjacent sides is a right angle.

**Cube numbers:** A number that can be obtained by multiplying together three of the same number.

**Cube root:** A number that makes another number when it is multiplied by itself twice. For example, the cube root of 8 is 2 since $2 \times 2 \times 2 = 8$.

**Cuboid:** A solid shape with six rectangular surfaces or four rectangular and two square surfaces.

**Decimal place:** the position of a digit after the decimal point, each successive position to the right having a denominator of an increased power of ten.

**Decrease:** To become less in quantity, size or intensity.

**Denominator:** The bottom number in a fraction.

**Edge:** The line where two faces or surfaces of a 3D shape meet.

**Enlargement:** A transformation in which the object is enlarged to form an image.

**Equally likely:** The outcomes of an event are equally likely when each has the same probability of occurring. For example, when tossing a fair coin the outcomes 'heads' and 'tails' are equally likely.

**Equation:** A relation in which two expressions are separated by an equals sign with one or more variables. An equation can be solved to find one or more answers, but it may not be true for all values of $x$.

**Equilateral (triangle):** A triangle in which all the sides are equal and all the angles are 60°.

**Equivalent fraction:** Any fraction that can be made equal to another fraction by cancelling. For example $\frac{9}{12} = \frac{3}{4}$.

**Expression:** A collection of numbers, letters, symbols and operators representing a number or amount; for example, $x^2 - 3x + 4$.

**Face:** The area on a 3D shape enclosed by edges.

**Factor:** A number that divides exactly (no remainder) into another number; for example, the factors of 12 are {1, 2, 3, 4, 6, 12}.

**Fair:** All outcomes are equally likely.

**Formula:** A mathematical rule, using numbers and letters, which shows a relationship between variables; for example, the conversion formula from temperatures in Fahrenheit to temperatures in Celsius is: $C = \frac{5}{9}(F - 32)$.

**Fraction:** A number that can be expressed as a proportion of two whole numbers.

**Frequency:** The number of times each value occurs.

**Frequency table:** A table that shows all the frequencies after all the data has been collected.

**Gradient:** The slope of a line; the vertical difference between the coordinates divided by the horizontal difference.

**Image:** The result of a reflection or other transformation of an object.

**Improper fraction:** A fraction that has a numerator greater than the denominator.

**Increase:** To make or to become greater in size, degree, frequency, etc; to grow or expand.

**Index (notation):** Expressing a number in terms of one or more of its factors, each expressed as a power.

**Intercept:** The point where a line cuts or crosses an axis.

**Interior angle:** The inside angle between two adjacent sides of a 2D shape, at a vertex.

**Isosceles (triangle):** A triangle in which two sides are equal and the angles opposite the equal sides are also equal.

**Litre:** A metric unit of volume that is a thousand cubic centimetres. It is equal to 1.76 British pints.

**Mean:** A number that is the average of a set of numbers.

**Median:** The middle value in a frequency distribution, below and above which lie values with equal total frequencies.

**Midpoint:** The point on a line that is at an equal distance from either end.

**Mixed number:** A number made up of a whole number and a fraction, for example $3\frac{3}{4}$.

**Mode:** The value in a range of values that has the highest frequency as determined statistically.

**Multiple:** Any member of the times table; for example multiples of 7 are 7, 14, 21, 28, etc.

***n*th term:** An expression in terms of $n$ where $n$ is the position of the term; it allows you to find any term in a sequence, without having to use a term-to-term rule.

**Numerator:** The top number in a fraction.

**Opposite (side):** The side that is the furthest away from a given angle, in a right-angled triangle.

**Outcome:** A possible result of an event in a probability experiment, such as the different scores when throwing a dice.

**Parallelogram:** A four sided shape where both pairs of opposite sides are parallel.

**Pattern:** Numbers or objects that are arranged to follow a rule.

**Percentage:** Any fraction or decimal expressed as an equivalent fraction with a denominator of 100 but written with a percentage sign (%), for example $0.4 = \frac{40}{100} = 40\%, \frac{4}{25} = \frac{16}{100} = 16\%$.

**Perimeter:** The curve or line enclosing a plane area; the length of this curve or line.

**Pictogram:** A frequency table where the frequency for each type of data is shown by a symbol.

**Pie chart:** A method of comparing discrete data. A circle is divided into sectors whose angles each represent a proportion of the whole sample.

**Polygon:** A closed 2D shape with straight sides.

**Power:** The number of times you use a number or expression in a repeated multiplication; it is written as a small raised number; for example, $2^2$ is 2 multiplied by itself, and $2^2 = 2 \times 2$ and $4^3 = 4 \times 4 \times 4$.

**Possibility:** The likelihood of something happening from a range of choices.

**Prime (number):** A number with only two factors, 1 and itself.

**Prism:** A 3D shape that has the same cross-section wherever it is cut perpendicular to its length.

**Proportion:** 1) A relationship that maintains a constant ratio between two variable quantities. 2) A relationship between four numbers or quantities in which the ratio of the first pair equals the ratio of the second pair.

**Quadratic:** Having terms involving one or two variables and a constant, such as $x^2 - 3$ or $y^2 + 2y + 4$, where the highest power of the variable is two.

**Quadrilateral:** A polygon having four sides.

**Random:** Chosen by chance, without selection; every item has an equal chance of being chosen.

**Range:** The difference between the highest and lowest values for a set of data.

**Ratio:** The ratio of A to B is a number found by dividing A by B. It is written as A:B. For example, the ratio of 1 m to 1 cm is written as 1 m : 1 cm = 100 : 1. Notice that the two quantities must both be in the same units if they are to be compared in this way.

**Reflection:** The image formed when a 2D shape is reflected in a mirror line or line of symmetry; the process of reflecting an object.

**Relative frequency:** An estimate for the theoretical probability.

**Rotation:** A turn about a central point, called the centre of rotation.

**Scale:** The number of squares that are used for each unit on an axis.

**Scale factor:** The ratio of the distance on the image to the distance it represents on the object; the number that tells you how much a shape is to be enlarged.

**Sequence:** A pattern of numbers that are related by a rule.

**Significant figure:** In the number 12 068, 1 is the first and most significant figure and 8 is the fifth and least significant figure. In 0.246, the first and most significant figure is 2. Zeros at the beginning or end of a number are not significant figures.

**Simplify:** To make an equation or expression easier to work with or understand by combining like terms or cancelling; for example, $4a - 2a + 5b + 2b = 2a + 7b$, $\frac{12}{18} = \frac{2}{3}$, $5:10 = 1:2$.

**Solve:** To work out the answer to a problem or to obtain the roots of an equation.

**Speed:** The rate at which an object moves; for example, the speed of the car was 40 miles per hour.

**Square:** A shape with four sides that are all the same length and four corners that are all right angles.

**Square numbers:** A number formed when any integer is multiplied by itself. For example, $3 \times 3 = 9$ so 9 is a square number.

**Square root:** A number that produces a specified quantity when multiplied by itself. For example, the square root of 16 is 4. Not all square roots are whole numbers. It uses the symbol $\sqrt{\phantom{x}}$, so $\sqrt{25} = 5$, and $\sqrt{7} = 2.645{,}751\ldots$

**Standard form:** A way of writing a number as $a \times 10^n$, where $1 \leq a < 10$ and $n$ is a positive or negative integer.

**Substitute:** Replace a variable in an expression with a number and work out the value; for example, if you substitute 4 for $t$ in $3t + 5$ the answer is 17 because $3 \times 4 + 5 = 17$.

**Symmetry:** An exact correspondence in position or form about a given point, line or plane.

**Term:** 1) A part of an expression, equation or formula. Terms are separated by + and – signs. 2) A number in a sequence or pattern.

**Term-to-term:** The rule that shows what to do to one term in a sequence to work out the next term.

**Translation:** A movement along, up, down or diagonally on a coordinate grid.

**Triangle:** An object, arrangement or flat shape with three straight sides and three angles.

**Triangle numbers:** All numbers that can be arranged in an equilateral triangular pattern.

**Vector:** A quantity such as velocity that has magnitude and acts in a specific direction.

**Vertex:** The point at which two lines meet in a 2D or 3D shape.

**Vertices:** The plural of vertex.

# Notes

Notes

# Answers

## 1.1 Multiples

1  45, 50, 55

2  42, 49, 56

3  even numbers

4  11, 22, 33, 44, 55, 66

5  20, 40, 60, 80, 100, 120, 140

6  21, 24, 27, 30, 33, 36, 39

7  Any multiple of 20, e.g. 20, 40, 60, 80, 100, …

8  a  A circle round 90, 93, 96, 99
   b  a square round 92, 96, 100
   c  96

9  a  4, 8, 12, 16, 20, 24    b  6, 12, 18, 24, 30, 36
   c  Any multiples of 12, e.g. 12, 24, …

10  a  15, 30, 45, 60, 75, 90
    b  10, 20, 30, 40, 50, 60, 70, 80, 90
    c  30, 60, 90           d  30

11  Any five multiples of 6, e.g. 6, 12, 18, 24, 30, …

## 1.2 Factors

1  1, 15, 3, 5

2  1, 2, 4, 5, 10, 20

3  50 and 20

4  a  1, 2, 4    b  1, 2, 3, 4, 6    c  1, 3, 5    d  1

5  a  1, 2, 3, 6, 9, 18    b  1, 3, 9, 27    c  1, 3, 9

6  a  1, 2, 4, 8, 16    b  1, 2, 4, 7, 14, 28    c  1, 2, 4

7  Any three from 1, 2, 5, 10

8  11

## 1.3 Prime numbers

1  a  10, 12, 14, 16, 18, 20 crossed out
   b  11, 13, 17, 19 circled

2  a  i  1, 21, 3, 7   ii  1, 22, 2, 11   iii  1, 23   iv  1, 5, 25
   b  23

3  23 and 29

4  89

5  Each number has a factor other than 1 and the number itself,
   a  e.g. even number so 2 is a factor
   b  e.g. 5 is a factor since the last digit is 5
   c  e.g. 9 is a factor since $9 \times 111 = 999$

6  $5 \times 7$ (or $7 \times 5$)

7  a  $2 \times 13$         b  $3 \times 11$         c  $5 \times 13$

8  a  12        b  70        c  110        d  125

9  a  20        b  140

10  a  $2 \times 3 \times 3$                b  $7 \times 11$
    c  $2 \times 3 \times 5$                d  $2 \times 2 \times 2 \times 2 \times 2$

## 1.4 Lowest common multiple and highest common factor

1  a  6, 12, 18, 24, 30, 36    b  4, 8, 12, 16, 20, 24
   c  12 and 24                d  12

2  a  10        b  20        c  30        d  10

3  12

4  50

5  a  1, 2, 3, 4, 6, 12    b  1, 2, 3, 6, 9, 18
   c  1, 2, 3, 6          d  6

6  a  1, 2, 5, 10    b  10

7  a  20        b  $\dfrac{2}{5}$

8  15

## 1 Problem solving

1  a  10 is a factor of 60 since $10 \times 6 = 60$
   b  8 is not a factor of 60; $7 \times 8 = 56$ so 7 rows would have 4 chairs left over.
   c  i  12 rows. The highest factor less than 14 is 12 because $12 \times 5 = 60$.
      ii  5 rows (12 chairs in each row)

2  a  $7 + 7$ and $3 + 11$   b  $5 + 19$, $7 + 17$ and $11 + 13$
   c  11 is a prime number and $385 = 35 \times 11$ or $385 \div 11 = 35$
   d  5 and 7 are prime factors; $5 \times 77 = 7 \times 55 = 385$
   e  1, 385, 35, 55 and 77

# ANSWERS TO CHAPTER 2: FRACTIONS, DECIMALS AND PERCENTAGES

## 2.1 Equivalent fractions

1 a $\frac{3}{6} = \frac{1}{2}$    b $\frac{12}{16} = \frac{3}{4}$    c $\frac{4}{12} = \frac{1}{3}$

2 a $\frac{1}{4}$    b $\frac{2}{5}$    c $\frac{2}{3}$    d $\frac{1}{2}$

3 a $\frac{1}{2} = \frac{3}{6}$   b $\frac{3}{5} = \frac{6}{10}$   c $\frac{1}{2} = \frac{6}{12}$   d $\frac{6}{8} = \frac{3}{4}$

   e $\frac{3}{2} = \frac{6}{4}$   f $\frac{2}{3} = \frac{8}{12}$   g $\frac{3}{4} = \frac{9}{12}$   h $\frac{5}{2} = \frac{15}{6}$

4 Possible answers include $\frac{3}{4}, \frac{6}{8}, \frac{12}{16}, \frac{15}{20}, \frac{18}{24}$

5 a $\frac{1}{2}$    b $\frac{1}{3}$    c $\frac{1}{4}$    d $\frac{3}{4}$

6 a $\frac{2}{3}$    b $\frac{5}{8}$    c $\frac{2}{5}$    d $\frac{3}{4}$

7 a 1    b $\frac{3}{5}$    c $\frac{1}{2}$    d $\frac{2}{5}$

   e 1    f $\frac{5}{4}$ or $1\frac{1}{4}$

8 b $1\frac{1}{4}$   c $2\frac{1}{4}$   d $1\frac{1}{2}$   e $3\frac{1}{2}$

   f $1\frac{3}{8}$   g $2\frac{1}{8}$   h $1\frac{7}{10}$   i $3\frac{3}{10}$

9 a $\frac{11}{2}$    b $\frac{13}{4}$    c $\frac{19}{8}$    d $\frac{17}{3}$

10 $\frac{2}{5}$   $\frac{1}{2}$   $\frac{2}{3}$   $\frac{3}{4}$

## 2.2 Arithmetic with fractions

1 a $\frac{6}{8}$      b $\frac{7}{8}$

2 a $\frac{3}{9}$      b $\frac{5}{9}$

3 a $\frac{1}{4} + \frac{2}{4} = \frac{3}{4}$    b $\frac{1}{8} + \frac{6}{8} = \frac{7}{8}$

   c $\frac{1}{6} + \frac{3}{6} = \frac{4}{6} = \frac{2}{3}$    d $\frac{5}{10} + \frac{1}{10} = \frac{6}{10} = \frac{3}{5}$

4 a $\frac{2}{4} = \frac{1}{2}$    b $\frac{3}{8}$    c $\frac{5}{8}$

   d $\frac{7}{8}$    e $\frac{6}{10} = \frac{3}{5}$    f $\frac{8}{10} = \frac{4}{5}$

5 a $\frac{1}{3}$      b $\frac{3}{5}$      c $\frac{5}{10} = \frac{1}{2}$

   d $\frac{8}{10} = \frac{4}{5}$    e $\frac{6}{8} = \frac{3}{4}$    f $\frac{2}{8} = \frac{1}{4}$

6 a $\frac{1}{4}$   b $\frac{3}{8}$   c $\frac{3}{8}$   d $\frac{3}{8}$   e $\frac{4}{10}$   f $\frac{2}{10}$

7 b $3\frac{1}{2}$   c $1\frac{1}{4}$   d $3\frac{3}{4}$   e 2   f $3\frac{1}{5}$

8 a $2\frac{1}{3}$   b $2\frac{2}{3}$   c $4\frac{1}{2}$   d $2\frac{2}{5}$   e $2\frac{1}{4}$   f $3\frac{1}{2}$

9

| × | $\frac{1}{4}$ | $\frac{1}{3}$ | $\frac{1}{2}$ | $\frac{2}{3}$ | $\frac{3}{4}$ |
|---|---|---|---|---|---|
| 3 | $\frac{3}{4}$ | 1 | $1\frac{1}{2}$ | 2 | $2\frac{1}{4}$ |
| 5 | $1\frac{1}{4}$ | $1\frac{2}{3}$ | $2\frac{1}{2}$ | $3\frac{1}{3}$ | $3\frac{3}{4}$ |

## 2.3 Fractions and decimals

1 a 0.25   b 0.75   c 2.5   d 5.25

2 a 0.3   b 0.9   c 1.1   d 2.7

3 a $\frac{2}{5}$   b $\frac{4}{5}$   c $\frac{3}{5}$   d $1\frac{1}{5}$

4 a $\frac{3}{20}$   b $\frac{7}{20}$   c $\frac{9}{20}$   d $3\frac{11}{20}$

5 a 0.375   b 0.625   c 0.875

6 A = 1.2 = $1\frac{1}{5}$     B = 1.8 = $1\frac{4}{5}$

   C = 2.3 = $2\frac{3}{10}$     D = 2.6 = $2\frac{3}{5}$

7 a 8.3   b 4.25   c 6.6   d 5.125

8 a $9\frac{7}{10}$   b $12\frac{4}{5}$   c $10\frac{3}{4}$   d $8\frac{1}{8}$

9 a $1\frac{1}{5} = 1.2$   b $1\frac{3}{5} = 1.6$   c $3\frac{3}{4} = 3.75$   d $5\frac{1}{4} = 5.25$

10 a 0.666…   b 0.111…   c 0.555…

11 $\frac{5}{8}$   $\frac{2}{3}$   $\frac{7}{10}$   $\frac{4}{5}$

## 2.4 Percentages

1 63% and 90%

2 a 0.44     b 0.4     c 0.04

3

| 0.5 | 0.35 | 0.77 | 0.6 | 0.8 | 0.06 |
|---|---|---|---|---|---|
| 50% | 35% | 77% | 60% | 80% | 6% |

4

| 25% | 70% | 5% | 75% | 90% | 15% |
|---|---|---|---|---|---|
| $\frac{1}{4}$ | $\frac{7}{10}$ | $\frac{1}{20}$ | $\frac{3}{4}$ | $\frac{9}{10}$ | $\frac{3}{20}$ |

5 15%

6 $\frac{1}{4} + \frac{1}{5} = \frac{9}{20}$

7 a £5   b £2   c £14   d £15   e £4   f £16

**8 a** $\frac{1}{5}$

**b i** £3 **ii** £8 **iii** 50p or £0.50
**iv** 11 kg **v** 7 cm **vi** 16 m

**9**

| £7.50 | £30 | £37.50 | £60 | £150 |
|---|---|---|---|---|
| £4 | £16 | £20 | £32 | £80 |

## 2.5 Working with fractions and percentages

**1 a** 10 kg **b** 12 m **c** 45 people
**d** 36 kg **e** 25 g **f** 21 m

**2 a** $\frac{1}{5}$ **b** 20%

**3 a i** $\frac{3}{4}$ **ii** $\frac{3}{10}$ **iii** $\frac{3}{5}$ **b i** 75% **ii** 30% **iii** 60%

**4 a** 0.32 **b i** £64 **ii** £38.40 **iii** £8 **iv** £1.92

**5 a i** 0.08 **ii** 0.45

**b**

| £2 | £3.20 | £4 | £19.20 |
|---|---|---|---|
| £11.25 | £18 | £22.50 | £108 |

**6 a** £6 **b** £66 **c** £54

**7 a** £32 **b** £112 **c** £48

**8 a** 6 kg **b** 126 kg **c** 114 kg

**9** e.g. 4% of £650 = 0.04 × £650 = £26; £650 + £26 = £676

## 2 Problem solving

**1 a** $\frac{3}{5}$ **b** 15 **c** 45%

**d** 60% of 90 = 54, **e** 54 − 18 = 36 more

**2 a i** £3 **ii** 25%

**b i** 40% of £80 = £32 so the sale price is £80 − £32 = £48
**ii** No. 40% of £210 = £84. The price with a 40% reduction is £210 − £84 = £126

# ANSWERS TO CHAPTER 3: WORKING WITH NUMBERS

## 3.1 Order of operations

**1 a** 11 **b** 20 **c** 8 **d** 12 **e** 11 **f** 29

**2 a** 12 **b** 2 **c** 5 **d** 50 **e** 25 **f** 6

**3 a** 10 **b** 4 **c** 1 **d** 2 **e** 12 **f** 5

**4 a** F 14 **b** F 6 **c** T **d** T **e** T **f** T
**g** F 5 **h** T **i** F 2 **j** F 2

**5** He did the subtraction first. He should do the multiplication first. The answer is 6.

**6** Two possible answers are 5 + 3 × 4 and 1 + 2 × 8.

**7 a** 25 **b** 2 **c** 60

## 3.2 Powers and roots

**1 a** 9 **b** 49 **c** 100

**2 a** 41 **b** 85

**3 a** 1.69 **b** 12.25 **c** 21.16

**4 a** 6 **b** 9 **c** 12 **d** 15 **e** 30 **f** 32

**5 a** 1.4 **b** 3.1 **c** 8.4

**6 a** 125 **b** 8 **c** 216 **d** 1000

**7 a** 1.331 **b** 17.576 **c** 531.441

**8 a** 2 **b** 5 **c** 10 **d** 8

**9 a** 16 **b** 256 **c** 64

**10** $3^4 = 81$ and $4^3 = 64$ and so $3^4$ is larger

**11** $3^2, 2^4, 4^3$

## 3.3 Rounding numbers

**1 a** 40 **b** 680 **c** 250

**2 a** 400 **b** 3600 **c** 3100

**3 a** 4000 **b** 6000 **c** 19 000

**4 a** 43 **b** 52 **c** 2 **d** 4
**e** 12 **f** 15 **g** 3 **h** 81
**i** 99 **j** 67 **k** 13 **l** 9

**5 b** 6.8 **c** 8.9 **d** 14.5 **e** 11.6
**f** 58.5 **g** 8.6 **h** 3.8

**6 a** 8.7 **b** 11.4 **c** 14.2

**7 a** 4.67 **b** 3.34 **c** 8.88 **d** 2.24
**e** 7.21 **f** 9.49

**8 a** 0.67 **b** 0.17 **c** 0.83

**9 a** 800 **b** 400 **c** 6000
**d** 7000 **e** 5000 **f** 60 000

## 3.4 Standard form

**1 a** 100 **b** 1000 **c** 100 000

**2 a** $10^3$ **b** $10^4$ **c** $10^6$

**3 a** 150 **b** 2400 **c** 38 000

**4 a** 73 **b** 580 **c** 2900

**5 a** 100 **b** 100 **c** 1000

**6 a** $3.2 \times 10^3$ **b** $6 \times 10^3$ **c** $1.23 \times 10^3$

**7 a** $4 \times 10^4$ **b** $7.3 \times 10^4$ **c** $2.6 \times 10^5$
**d** $8.03 \times 10^5$ **e** $7 \times 10^6$ **f** $4.8 \times 10^6$

**8 a** 900 **b** 920 **c** 300 000
**d** 1 800 000 **e** 475 000 **f** 14 100

**9** B C A D

**10 a** $5.1 \times 10^5$ **b** $3.3 \times 10^5$

## 3.5 Units

**1 a** 25 **b** 45 **c** 60 **d** 50

**2** 1.65 m

**3** 74 kg

**4** 45 minutes

**5 a** 2000 **b** 1500 **c** 250

**6 a** cm **b** ml or l **c** kg
**d** km **e** g **f** m

**7 a** 9     **b** 400     **c** 2000

   **d** 50     **e** 500     **f** 53

**8 a** 3500    **b** 4650    **c** 350

**9 a** 6500    **b** 2850    **c** 610

**10 a** 18 : 20   **b** 14 : 05

## 3 Problem solving

**1 a** 5 and 4 in that order

   **b** One pair is 2 and 4. The other pair is 3 and 5. The order does not matter.

   **c** One pair is 2 and 3. The other pair is 4 and 5. The answer is 26.

   **d** One pair is 2 and 5. The other pair is 3 and 4 to make 49

**2 a** 126 000, 288 000      **b** 90 000 , 100 000, 300 000

   **c** $5.02 \times 10^5$

# ANSWERS TO CHAPTER 4: SEQUENCES OF NUMBERS

## 4.1 Describing a sequence

**1 a** 16     **b** 19     **c** 26

   **d** 13     **e** 51     **f** 40

**2 a** 25, 28, 31, 34

   **b** Add 3 to the last number to get the next

**3 a** 27     **b** 52     **c** 64

   **d** 26     **e** 56     **f** 24

**4 a** add 5    **b** add 7    **c** subtract 2    **d** subtract 4

**5 a** 13 and 18     **b** 33

**6 a** multiply by 2    **b** multiply by 3

   **c** multiply by 5    **d** multiply by 4

**7** 10, 20, 40

**8 a** 52      **b** subtract 12

## 4.2 Recognising sequences

**1 a** 16, 25, 36     **b** 64, 100

**2** 25, 49, 81, 121

**3 a** 5, 7, 9, 11   **b** odd numbers adding 2 each time

**4** 64, 125, 216

**5** $10 \times 10 \times 10 = 1000$

**6 a** $1 + 2 + 3 + 4 = 10$     **b** 15    **c** 21     **d** 55

**7** $8 \times 8 = 64$ and $4 \times 4 \times 4 = 64$

## 4.3 The $n$th term of a sequence

**1 a** 5     **b** 9     **c** 1     **d** 14

**2 a** 6     **b** 12     **c** 15     **d** 30

**3 a** 8     **b** 13     **c** 3     **d** 10

**4** bottom line 6, 8, 10, 12, 14

**5** bottom line 4, 6, 8, 10, 12

**6 a** 1     **b** 10     **c** 13     **d** 28

**7** 7, 9, 11, 13

**8** 1, 4, 7, 10

**9** $2n - 1$      2, 4, 6, 8, 10, …

   $2n$      1, 3, 5, 7, 9, …

   $2n + 1$      4, 6, 8, 10, 12, …

   $2(n + 1)$      3, 5, 7, 9, 11, …

**10 a** 11     **b** 23     **c** 35

**11 a** 11, 21, 31, 41, 51     **b** 101

## 4 Problem solving

**1 a** 1, 3, 5, 7, 9     **b i** 4    **ii** 8    **iii** 12    **iv** 16

   **c** The first four multiples of 4

   **d** 1, 3, 6, 10, 15    **e i** 4    **ii** 9    **iii** 16    **iv** 25

   **f** square numbers      **g i** 36   **ii** 100

**2 a** dots 4, 6, 8, 10      **b** 14

   **c** lines 4, 7, 10, 13      **d** 22

   **e** If $n = 4$ then $3 \times 4 + 1 = 13$   **f** 61

# ANSWERS TO CHAPTER 5: COORDINATES AND GRAPHS

## 5.1 Coordinates

**1 a** A (3, 2)   C (−3, −4)   D (3, −4)

   **b** Point (0, 2) marked on the diagram

   **c** (3, −1) (−3, −1) (0, −4)

**2 a and b**

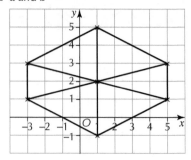

   **c** (1, 2)

**3 a and b**

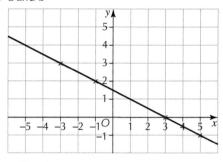

**c** (0, 1.5)

**4 a–d**

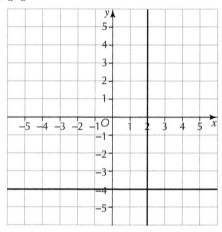

**e** (2, −4)

## 5.2 The equation of a straight line

**1 a** B $x = 5$, C $x = −4$    **b** E $y = −3$, F $y = 3$

**2 a i** 2    **ii** 1    **iii** −2

**b**

| $x$ | 2 | 0 | −2 | −4 |
|---|---|---|---|---|
| $y = x + 3$ | 5 | 3 | 1 | −1 |

**c and d**

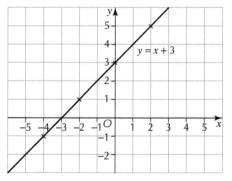

**e** Possible answers are (1, 4) or (−1, 2) or (−3, 0) or (−5, −2)

**3 a i** −4    **ii** −6    **iii** −10

**b**

| $x$ | −2 | −1 | 0 | 1 | 2 |
|---|---|---|---|---|---|
| $y = 2x$ | −4 | −2 | 0 | 2 | 4 |

**c and d**

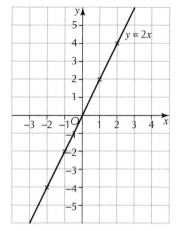

**e** Possible answers are (0.5, 1) or (1.5, 3) or (−1.5, −3)

**4 a i** 3    **ii** 1    **iii** 4    **iv** −1

**b**

| $x$ | −1 | 0 | 1 | 2 | 3 | 4 | 5 |
|---|---|---|---|---|---|---|---|
| $y$ | 5 | 4 | 3 | 2 | 1 | 0 | −1 |

**c**

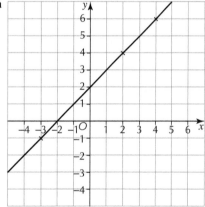

## 5.3 Intercept and gradient

**1 a**

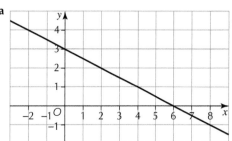

**b** (0, 2) and (−2, 0)

**2 a**

**b** Circle (4, 1) and (8, −1)

**3 a** 4  **b** $\frac{1}{3}$

**4 a** 3  **b** $\frac{1}{2}$  **c** $\frac{1}{5}$

**5 a**

| $x$ | −2 | −1 | 0 | 1 | 2 | 3 |
|---|---|---|---|---|---|---|
| $y = 2x + 4$ | 0 | 2 | 4 | 6 | 8 | 10 |

**b**

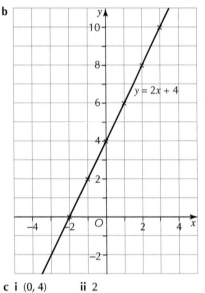

**c i** (0, 4)  **ii** 2

## 5.4 Quadratic graphs

**1 a**

| $x$ | 0 | 1 | 2 | 3 | 4 | 5 |
|---|---|---|---|---|---|---|
| $y = x^2$ | 0 | 1 | 4 | 9 | 16 | 25 |

**b and**

**c**

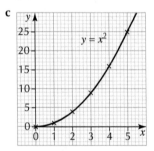

**2 a**

| $x$ | 0 | 1 | 2 | 3 | 4 |
|---|---|---|---|---|---|
| $y = x^2 + 4$ | 4 | 5 | 8 | 13 | 20 |

**b and c**

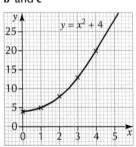

**3 a**

| $x$ | 0 | 1 | 2 | 3 | 4 | 5 |
|---|---|---|---|---|---|---|
| $y = x^2 - 10$ | −10 | −9 | −6 | −1 | 6 | 15 |

**b**

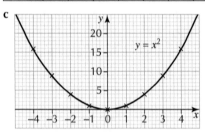

**4 a i** 4  **ii** 16

**b**

| $x$ | −4 | −3 | −2 | −1 | 0 | 1 | 2 | 3 | 4 |
|---|---|---|---|---|---|---|---|---|---|
| $y = x^2$ | 16 | 9 | 4 | 1 | 0 | 1 | 4 | 9 | 16 |

**c**

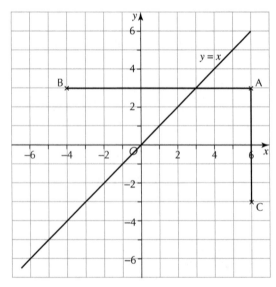

## 5 Problem solving

**1 a and e**

**b** (1, 3)  **c** (6, 0)  **d** $y = 3$  **f** (3, 3)  **g** (−4, −3)

**2 a**

| $x$ | 0 | 1 | 2 | 3 |
|---|---|---|---|---|
| $3x$ | 0 | 3 | 6 | 9 |

**b, e and i**

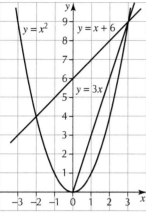

$y = x^2$   $y = x + 6$   $y = 3x$

**c** 3

**d**

| $x$ | −3 | −2 | −1 | 0 | 1 | 2 | 3 |
|---|---|---|---|---|---|---|---|
| $x + 6$ | 3 | 4 | 5 | 6 | 7 | 8 | 9 |

**f** 1        **g** (0, 6)

**h**

| $x$ | −3 | −2 | −1 | 0 | 1 | 2 | 3 |
|---|---|---|---|---|---|---|---|
| $x^2$ | 9 | 4 | 1 | 0 | 1 | 4 | 9 |

**j** (3, 9)

**6**
4(x + 2) ——— 8x + 2
2(x + 4) ——— 8x + 4
8(x + 1) ——— 8x + 8
2(4x + 1) ——— 2x + 8
4(2x + 1) ——— 4x + 8

**7 a** $2(c + 5)$   **b** $3(x − 3)$   **c** $2(w + 5)$   **d** $2(4n − 1)$

**8 a** $4(a + b)$   **b** $6(x − 2y)$   **c** $10(2s + 3t)$   **d** $4(a + 2b − 3c)$

## 6.3 Solving equations

**1 a** $x = 15$      **b** $y = 7$

**2 a** $t = 4$      **b** $y = 6$

**3 a** $g = 11$      **b** $r = 12$

**4 a** $p = 6$      **b** $T = 18$

**5 a** $x = 4$      **b** $y = 8$

**6 a** $y = 4$   **b** $x = 8$   **c** $w = 7$   **d** $g = 4$   **e** $m = 30$

## 6 Problem solving

**1 a** $L + L + L + L = 4L$      **b** $6L$
   **c** $(6L + 12)$ or $6(L + 2)$

**2 a** 12      **b** 9, 18      **c** 7, 24      **d** 15, 20

# ANSWERS TO CHAPTER 6: EXPRESSIONS AND EQUATIONS

## 6.1 Substituting into formulae

**1 b** $R + 4$        **c** $2R$        **d** $\frac{1}{2}R$ or $R ÷ 2$

**2 a** 15      **b** 1      **c** 3
   **d** −2      **e** 12      **f** 30

**3 b** $20 − 12 = 8$      **c** $40 + 18 = 58$      **d** $2 + 3 = 5$
   **e** $2 × 10 = 20$      **f** $5 × 2 = 10$

**4 a** 12 m      **b** 18 m

**5 a** 12 cm      **b** 27 m

**6 a** £50      **b** £90

**7 a** 20 cm²      **b** 80 cm²

## 6.2 Simplifying expressions

**1 a** $5x$      **b** $2y$      **c** $8p$      **d** $5t$

**2 a** $2t + 3$   **b** $4x$   **c** $2 + 3k$   **d** $4n − 3$

**3** $3t + 4$

**4 a** $2x + 2$   **b** $3y − 6$   **c** $4a + 12$   **d** $10a − 5$

**5 a** $4x + 4y$   **b** $3a + 6b$

# ANSWERS TO CHAPTER 7: RATIO AND PROPORTION

## 7.1 Ratio notation

**1** 16

**2 a** 7 : 4      **b** 4 : 7

**3** 2 : 1

**4 a** 3      **b** 3

**5 b** 3 : 1      **c** 4 : 1      **d** 3 : 4
   **e** 3 : 2      **f** 2 : 5      **g** 2 : 3

**6 a** 1 : 4      **b** 2 : 1      **c** 3 : 2

**7 a** 2 : 1      **b** 8 : 5

**8 a** 1 : 10      **b** 4 : 1

**9 a** true   **b** false   **c** true   **d** true

**10** 3 : 1

**11** 3 : 2

## 7.2 Dividing using a ratio

**1 a** 5      **b** 10      **c** 1 : 2

**2 a** 30      **b** 3 : 1

**3 a** 5      **b** 25

**4 a** 12      **b** 6

**5** 5 : 1

**6 a** 8      **b** 30

**7 a** 50 g      **b** 20 g

## 7.3 Proportion

1. **a** 18 km     **b** 5 hours
2. **a** 42 hours     **b** 133 hours
3. **a** £9.40     **b** £329
4. **a** £28.75     **b** 40 litres
5. **a** 16 g     **b** 90 g     **c** 100 ml
6. **a** 640 cm     **b** 6.4 m
7. **a** 32     **b** 12.5
8. **a** $1.30     **b** $123.50
9. **a** 63     **b** 150

## 7.4 Ratios and fractions

1. **a** 15     **b** 20     **c** $\frac{3}{4}$
2. **a** 2 : 1     **b** $\frac{2}{3}$
3. **a** 8 and 24     **b** $\frac{1}{4}$
4. $\frac{2}{5}$
5. 3 : 2
6. **a** 4 : 5     **b** $\frac{5}{9}$
7. **a** 3 : 4     **b** $\frac{3}{4}$     **c** $\frac{4}{3}$
8. **a** 7 : 3     **b** $\frac{3}{10}$     **c** $\frac{3}{7}$

## 7 Problem solving

1. **a** 20     **b** 30p     **c** £1.20     **d** 2 : 3
   **e** $\frac{2}{5}$     **f** 11p     **g** £1.10
   **h** 500 g bag costs £1.20 per 100 g and 380 g bag cost £1.10 per 100 g, so 380 g better value
2. **a** 16 km/l     **b** 960 km     **c** £15.84     **d** €22
   **e** £14.40     **f** 120 miles     **g** 10 miles/l
   **h** 45 miles/gallon

# ANSWERS TO CHAPTER 8: PERCENTAGES

## 8.1 One number as a percentage of another

1. **a** $\frac{14}{100}$     **b** $\frac{28}{100}$     **c** $\frac{35}{100}$     **d** $\frac{70}{100}$
2. **a** 41%     **b** 82%     **c** 88%     **d** 30%
3. **a** 15%     **b** 78%     **c** 3%
4. **a** 67%     **b** 44%     **c** 78%     **d** 43%
5. **a** 65%     **b** 14     **c** 35%
6. **a** 65%     **b** 20%
7. **a** 85%     **b** 15%
8. Ali 48%, Beth 32%, Carl 20%
9. **a** 79%     **b** 21%

## 8.2 Comparisons using percentages

1. 72%, 66%, 65%
2. **a** 80%     **b** 70%     **c** maths
3. **a** 36
   **b** 40% of girls have a pet which is less than 45%
4. $\frac{1}{4}$ = 25%, $\frac{1}{3}$ = 33%, 25% < 30% < 33%
5. **a** Dan 12.5%, Alice 10%     **b** Alice
6. **a** 71%     **b** Betaville is 66% which is less than 71%
7. **a** 79%     **b** B is 84% which is greater than 79%

## 8.3 Percentage change

1. **a** 150     **b** 750
2. **a** 24 cm     **b** 16 cm
3. **a** £0.78     **b** £16.38 an hour
4. **a** £64     **b** £384
5. **a** 1960     **b** 3640
6. **a** £39     **b** £91
7.

| Item | Original price | Reduction | Sale price |
|------|------|------|------|
| Chair | £240 | £60 | £180 |
| Table | £600 | £150 | £450 |
| Bed | £840 | £210 | £630 |
| Cupboard | £360 | £90 | £270 |

8. £12 000

## 8 Problem solving

1. **a** 66 ÷ 80 = 0.825 = 82.5%     **b** 70% and 68%
   **c** Test A; A is 70%, B is 45% and C is 58%
   **d i** 350     **ii** 64%
   **e** Ali 72%, Emily 64%, Fran 56.3% so only Ali gets a Distinction.
2. **a** 16%     **b** £500     **c** £812.50     **d** 44%     **e** £625

# ANSWERS TO CHAPTER 9: ANGLES AND POLYGONS

## 9.1 Points and lines

1. $a = 90°$     $b = 135°$     $c = 56°$
2. $a = 120°$     $b = 205°$     $c = 120°$
3. **a** 30°     **b** 150°

**4** They add up to 355°, not 360°

**5 a** vertically opposite    **b** alternate
  **c** corresponding

**6** $a = 75°$ $b = 105°$ $c = 155°$ $d = 155°$

## 9.2 Triangles

**1 a** 30°    **b** 35°    **c** 110°    **d** 20°

**2** 45°

**3** $x = 60°$, $y = 30°$

**4** Third angle is 180° − 64° − 52° = 64°, so two angles
  are 64°, therefore isosceles.

**5** R = 57°, P = 66°

**6** 180° − 130° = 50°; 50° ÷ 2 = 25°; each angle is 25°

**7** $a = 60°$, $b = 75°$

## 9.3 Quadrilaterals and other polygons

**1** 360°

**2** $a = 115°$, $b = 75°$, $c = 45°$, $d = 215°$

**3 a** 70°    **b** 110°

**4 a** kite    **b** 85°    **c** 40°

**5 a** trapezium    **b** 108°    **c** 125°

**6 a** hexagon    **b** 720°

**7 a** pentagon    **b** 150°    **c** 540°

## 9.4 Shapes on coordinate axes

**1 a, b**

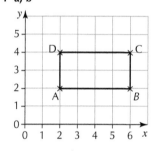

  **c** (2, 4)    **d** $x = 4$ and $y = 3$

**2 a, b, c**

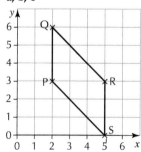

  **d** (5, 0)    **e** 45° and 135°

**3 a, b, c**

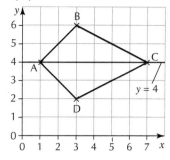

  **d** (3, 2)    **e** 90°

**4 a, b**

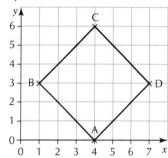

  **c** B(1, 3), D(7, 3)

**5 a, b**

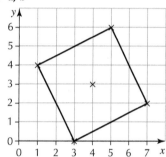

**6 a, b, c**

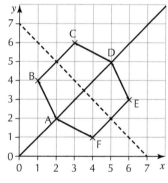

  **d** $x + y = 7$

## 9 Problem solving

**1 a**

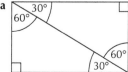

**b**

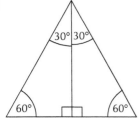

**c**

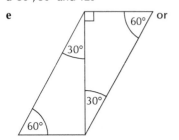

**d** 30°, 30° and 120°

**e**

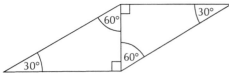

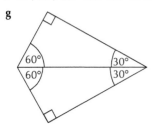

**f** 60°, 60°, 120° and 120° or 30°, 30°, 150° and 150°

**g**

**h** 60°, 90°, 90° and 120°

**2 a** 3

   **b** 3 angles = 360° so 1 angle = 360° ÷ 3 = 120°

   **c** 4

   **d** all 3 sides are the same length because they are sides of a regular hexagon.

   **e** rhombus    **f** 60°, 60°, 120° and 120°

   **g** 135°; rectangle angle + triangle angle = 90 + 45 = 135°

## 10.1 Perimeter and area

**1 a** 32 cm          **b** 26 cm

**2 a** 20 cm          **b** 22 m

**3 a** 21 cm²        **b** 30 m²

**4 a** 5 cm           **b** 25 cm²

**5 a** 50 cm²        **b** 25 cm²

**6 a** 14 cm²        **b** 24 m²

**7 a** 28 m           **b** 38 cm

**8 a** 40 m²         **b** 70 cm²

**9 a** 30 cm         **b** 42 cm²

**10 a** 30 cm       **b** 36 cm²

## 10.2 Circles

**1 a** chord         **b** tangent

**2 a** 16 cm         **b** 50 cm

**3** 63 cm

**4** 32 cm

**5 a** 75 cm   **b** 19 cm   **c** 53 cm   **d** 21 cm

**6** $\pi \times 5^2 = 79$ cm²

**7 a** 452 cm²  **b** 28 cm²  **c** 227 cm²  **d** 36 cm²

**8 a** 24 cm   **b** 88 cm²

## 10.3 Solids

**1 a** 6          **b** 12         **c** 8

**2 a** 5          **b** 9          **c** 6

**3 a** 20 cm²    **b** 12 cm²

**4** 32 cm²

**5 a** 9 cm²     **b** 63 cm³

**6 a** 180 cm³   **b** 288 cm³

**7 a i** 8    **ii** 18    **iii** 12    **b** 270 cm³

**8 a** 64 cm²   **b** 512 cm³

## 10 Problem solving

**1 a** 20 cm    **b** 24 cm²    **c** 28 cm     **d** 48 cm²

**e**

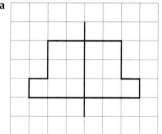

**f i** 32 cm      **ii** 48 cm²

**g** Possible answers include 3 cm by 8 cm **or** 2 cm by 12 cm **or** $1\frac{1}{2}$ cm by 16 cm **or** 1 cm by 24 cm

**h** Following on from part **g**, possible answers are 22 cm or 28 cm or 35 cm or 50 cm

**i** 25 cm²

**2 a** 9 cm²      **b** 54 cm²      **c** 27 cm³
  **d** 36 cm       **e** 54 cm³      **f** 18 cm²
  **g** 90 cm²     **h** 48 cm       **i** 144 cm

# ANSWERS TO CHAPTER 11: TRANSFORMATIONS

## 11.1 Reflections

**1 a**

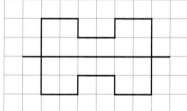

**b**

**2 a**

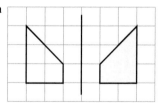

**b**

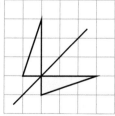

**3**

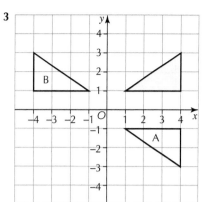

**4**

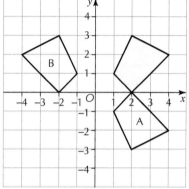

**5**

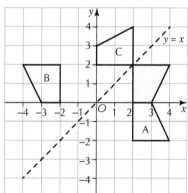

**6**

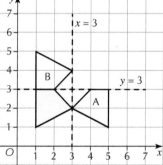

**7**

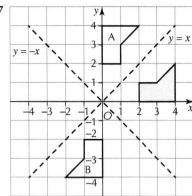

## 11.2 Translations and rotations

**1 a, b**

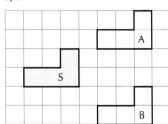

**c** $\begin{pmatrix} 0 \\ 4 \end{pmatrix}$

**2 a** $\begin{pmatrix} 5 \\ 3 \end{pmatrix}$     **b** $\begin{pmatrix} -5 \\ 0 \end{pmatrix}$     **c** $\begin{pmatrix} 5 \\ -3 \end{pmatrix}$

**3**

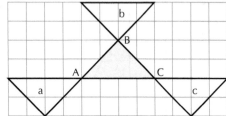

**4**

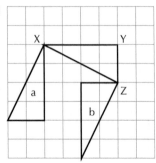

**5**

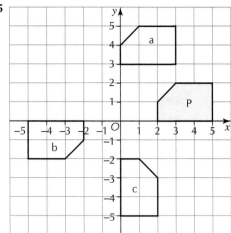

**6**

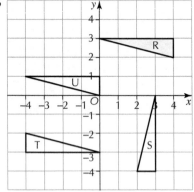

**7 a, b, d**

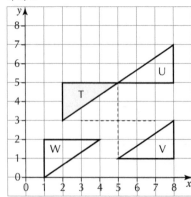

**c** $\begin{pmatrix} 0 \\ -4 \end{pmatrix}$

## 11.3 Enlargements

**1 a** 3     **b** 18 cm

**2 a** 4     **b** 48 cm

**3** 180 cm

**4**

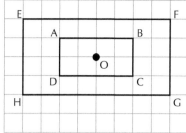

**5**

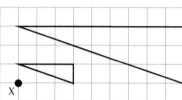

**6**

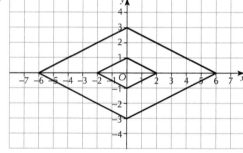

**7**

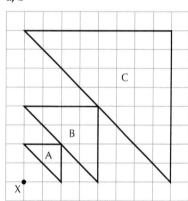

**8**

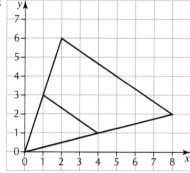

## 11 Problem solving

**1 a – f, h**

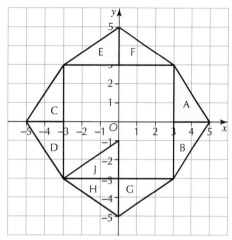

**g** rotation 90° clockwise, centre (0, 0)

**i** Reflection in the line $y = -3$

**2 a, b**

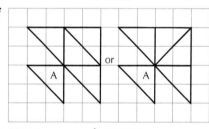

**c** 12 cm      **d** 24 cm      **e** 4

**f**

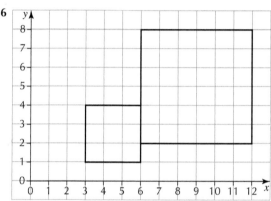

**g** 36 cm²      **h** 144 cm²

# ANSWERS TO CHAPTER 12: PROBABILITY

## 12.1 Relative frequency

**1 a** $\dfrac{13}{20}$      **b** $\dfrac{7}{20}$

**2 a** $\dfrac{9}{50}$      **b** $\dfrac{23}{50}$

3 a 200    b i 77%    ii 7%    iii 93%

4 a The frequencies for 4, 5, 6 are 2, 5, 2
  b i 0.2    ii 0.1    iii 0.6

5 a $\frac{573}{1000}$ or 0.573 or 57.3%    b 40

6 a $\frac{120}{200}$ or $\frac{3}{5}$ or 0.6 or 60%    b $\frac{20}{120}$ or $\frac{1}{6}$

  c $\frac{20}{32}$ or $\frac{5}{8}$ or 0.625 or 62.5%

## 12.2 Calculating probabilities

1

2 a $\frac{1}{6}$    b $\frac{2}{3}$    c $\frac{1}{2}$

3 a $\frac{3}{5}$    b $\frac{2}{5}$    c $\frac{9}{10}$

4 a i $\frac{2}{5}$    ii $\frac{1}{2}$    iii $\frac{9}{10}$    b i $\frac{4}{9}$    ii $\frac{5}{9}$    iii 1

5 a $\frac{2}{3}$    b $\frac{1}{3}$    c $\frac{1}{6}$

6 6

7 a i $\frac{1}{5}$    ii $\frac{12}{25}$    iii $\frac{8}{25}$    b 24

## 12.3 Possibility spaces

1 a AY, BX, BY, CX, CY    b AX, AY, AZ, CX, CY, CZ

2 a

| 1H | 2H | 3H | 4H | 5H | 6H |
|----|----|----|----|----|----|
| 1T | 2T | 3T | 4T | 5T | 6T |

b 12    c i $\frac{1}{12}$    ii $\frac{3}{12} = \frac{1}{4}$

3 a

| RR | YR | BR |
|----|----|----|
| RY | YY | BY |
| RB | YB | BB |

b 9    c i $\frac{1}{9}$    ii $\frac{3}{9} = \frac{1}{3}$    iii $\frac{6}{9} = \frac{2}{3}$

4 a 36    b $\frac{6}{36} = \frac{1}{6}$

c i $\frac{1}{36}$    ii $\frac{6}{36} = \frac{1}{6}$    iii $\frac{9}{36} = \frac{1}{4}$    iv $\frac{11}{36}$

5 a

| 7 | 8 | 9 | 10 | 11 | 12 |
|---|---|---|----|----|----|
| 6 | 7 | 8 | 9 | 10 | 11 |
| 5 | 6 | 7 | 8 | 9 | 10 |
| 4 | 5 | 6 | 7 | 8 | 9 |
| 3 | 4 | 5 | 6 | 7 | 8 |
| 2 | 3 | 4 | 5 | 6 | 7 |

b 7    c 2 or 12

d i $\frac{1}{36}$    ii $\frac{2}{36} = \frac{1}{18}$    iii $\frac{5}{36}$    iv $\frac{6}{36} = \frac{1}{6}$

## 12 Problem solving

1 a i 0.3    ii 0.15    iii 0.55

b

| Green | Yellow | Red | Total |
|-------|--------|-----|-------|
| 13 | 9 | 28 | 50 |

c i 0.26    ii 0.18    iii 0.56

d i 0.2    ii 0.2    iii 0.6

e No. They do not have to be exactly the same. They are quite similar. No convincing evidence it is unfair.

2 a

| HH | TH |
|----|----|
| HT | TT |

b i $\frac{1}{4}$    ii $\frac{1}{2}$

c

| HHH | HTH | THH | TTH |
|-----|-----|-----|-----|
| HHT | HTT | THT | TTT |

d i $\frac{1}{8}$    ii $\frac{3}{8}$    iii $\frac{3}{8}$

e This is a possible answer. The order of the columns could be different.

**Coins 1, 2 and 3**

| | | HHH | HHT | HTH | HTT | THH | THT | TTH | TTT |
|---|---|-----|-----|-----|-----|-----|-----|-----|-----|
| Coin 4 | Head (H) | HHHH | HHTH | HTHH | HTTH | THHH | THTH | TTHH | TTTH |
| | Tail (T) | HHHT | HHTT | HTHT | HTTT | THHT | THTT | TTHT | TTTT |

f i $\frac{1}{16}$    ii $\frac{4}{16} = \frac{1}{4}$    iii $\frac{6}{16} = \frac{3}{8}$

# ANSWERS TO CHAPTER 13: STATISTICS

## 13.1 Tables, charts and diagrams

**1 a** B     **b** C     **c** £54

**2 a** 25

**b**

**Telephone calls to helpline**

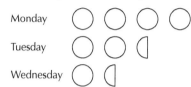

Monday

Tuesday

Wednesday

**3 a** Belgium     **b** 6 million     **c** 48 million

**4 a** January and February     **b** May     **c** 7

**5 a** 9     **b** 4     **c** 5

## 13.2 Calculating statistics

**1 a** football     **b** rugby

**2 a** 17 years     **b** 15 years     **c** 9 years

**3 a i** £9.30    **ii** £6.30    **b i** £10.15    **ii** £6.30

**4 a**

| 3 | 4 | 5 | 6 | 7 |
|---|---|---|---|---|
| 7 | 2 | 3 | 0 | 1 |

**b i** 1    **ii** 2    **iii** 7

**5 a** 8 minutes     **b** 4.8 minutes

**6 a** 23     **b** 37     **c** 32

**7 a i** 6 hours    **ii** 4 hours    **iii** 5 hours
   **b** 21 hours

## 13 Problem solving

**1 a** 1900 Kj     **b** Crispbread     **c** 300 Kj

**d**

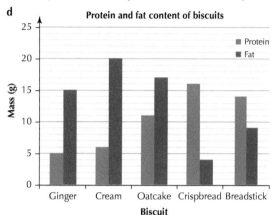

**e** 11 g     **f** 16 g     **g i** C    **ii** E    **iii** B

**2 a** 14 minutes     **b** 9 minutes     **c** 19 minutes

**d** 17 minutes

**e**

| 8 | 9 | 10 | 11 | 12 | 13 | 14 |
|---|---|----|----|----|----|----|
| 6 | 5 | 2  | 6  | 5  | 5  | 11 |

**f i** 6 minutes    **ii** 14 minutes    **iii** 12 minutes

**g** 11.45 minutes

**h** Task 5. The mean time is 8.75 minutes, which is 2.7 minutes less than task 4.

GERMAN FOR ENGINEERS

MODERNES FLIESSBAND IN DEN NSU WERKEN
Fertigstellung eines NSU Prinz 1000

# GERMAN FOR ENGINEERS

by

## NORMAN C. REEVES, B.A.

*Head of the Department of General Studies*
*Chance Technical College, Smethwick*

LONDON
SIR ISAAC PITMAN & SONS LTD.

*First published* 1965

SIR ISAAC PITMAN & SONS Ltd.
PITMAN HOUSE, PARKER STREET, KINGSWAY, LONDON, W.C.2
THE PITMAN PRESS, BATH
PITMAN HOUSE, BOUVERIE STREET, CARLTON, MELBOURNE
20–25 BECKETT'S BUILDINGS, PRESIDENT STREET, JOHANNESBURG

ASSOCIATED COMPANIES
PITMAN MEDICAL PUBLISHING COMPANY Ltd.
46 CHARLOTTE STREET, LONDON, W.I
PITMAN PUBLISHING CORPORATION
20 EAST 46TH STREET, NEW YORK, N.Y. 10017
SIR ISAAC PITMAN & SONS (CANADA) Ltd.
(INCORPORATING THE COMMERCIAL TEXT BOOK COMPANY
PITMAN HOUSE, 381–383 CHURCH STREET, TORONTO

MADE IN GREAT BRITAIN AT THE PITMAN PRESS, BATH
F5—(F.145)

# PREFACE

THE engineer of today has to be cosmopolitan in outlook. He is in competition with the engineers of other countries, and all are making things for a world market. Intelligent engineers are always anxious to learn, to pick up useful tips from others engaged in the same work.

The British engineer has had a good start over engineers of other lands, because the Industrial Revolution began in England, and he has been able to teach others much that he learnt. The foreign engineers have been eager and clever pupils. They now have something to teach the British.

German engineers are known to be good, and they have never been slow to learn from their rivals. We shall be wise to follow their example.

The *Object* of this book is to teach Engineers to READ German engineering texts.

No previous knowledge of German is presumed, but some understanding of grammar is of considerable advantage to the student. Allowances, however, have been made for those who have forgotten the grammar they learnt at school.

The book is not intended to provide the latest engineering information or to supply a comprehensive vocabulary of engineering terms.

The *Method* of the book is to present—
- (*a*) a passage of German,
- (*b*) its vocabulary,
- (*c*) an explanation or translation of the idioms or difficult passages,
- (*d*) a grammatical explanation, with rules.

The German passages are graded, becoming gradually harder. Effort has been made to include in each German passage examples of the constructions dealt with in the grammar section which follows it, but all the constructions are not dealt with systematically as soon as they occur. This does not hinder understanding of the passage, becàuse where translation may be difficult for the student a translation is given.

When all the common grammar has been handled, passages selected from a variety of German engineering books, pamphlets and journals are given. Each of these is supplied with all necessary elucidations.

The book is completed with a summary of German grammar and a comprehensive vocabulary.

The author wishes to acknowledge the valuable help he has received from Dr. Karl Gordon and Herr Ing. Franz Müller, both of Salzburg, and from Mr. W. F. Walker and the engineering staff of the Chance Technical College, Smethwick.

# CONTENTS

vii

# 1

# DIE WERKSTÄTTE

Hans ist ein Arbeiter. Er ist Mechaniker. Er arbeitet in einer Werkstätte. Die Werkstätte ist in einer Fabrik. Die Fabrik ist groß, sehr groß. Viele Männer und Frauen arbeiten dort. Sie machen Schrauben, Schrauben für Holz und Schrauben für Metall. In der Werkstätte sind viele Maschinen. Sie sind nicht sehr groß aber sie sind stark. Sie machen viel Lärm. Hans ist nicht allein in der Werkstätte. Fünf andere Männer sind auch dort. Einer ist der Werkmeister, Herr Braun. Er ist der Werkführer von Hans und den anderen. Er spricht laut, er schreit: die Maschinen machen so viel Lärm.

## VOCABULARY

| | | | |
|---|---|---|---|
| der Arbeiter | the worker, work-man | ist | is (from sein to be) |
| der Mechaniker | the mechanic | sind | are (from sein to be) |
| die Werkstätte | the workshop | arbeitet | works (from arbeiten to work) |
| die Fabrik | the factory | | |
| der Mann, Männer | the man, men | arbeiten | work |
| | | machen | make |
| die Frau, Frauen | the woman, women | spricht | speaks (from sprechen to speak) |
| die Schraube, Schrauben | the screw, screws | schreit | shouts (from schreien to shout) |
| das Holz | the wood | | |
| das Metall | the metal | ein, einer | a, one |
| die Maschine, Maschinen | the machine, machines | der, die, das | the |
| | | viel | much |
| der Lärm | the noise | viele | many |
| der Werk-meister | the foreman | er, sie, es | see Grammar below |
| der Werkführer | the supervisor | allein | alone |

1

| groß | big, large | und | and |
| stark | strong, robust | dort | there |
| fünf | five (*for numerals* | für | for |
|  | *see p.* 55) | nicht | not |
| laut | loud, loudly | aber | but |
| anderen | others | auch | also |
| in | in | von | of, from |
| sehr | very |  |  |

## GRAMMAR

### GENDER OF NOUNS: PRONOUNS

German nouns for things, as well as persons and animals with sex differences, have "gender," i.e. they may be referred to by pronouns as *he, she* and *it*.

Thus *the pencil* is **der Bleistift,** and the pronoun used for it is **er** (*he*);

the pen is **die Feder,** and the pronoun used for it is **sie** (*she*);
*the paper* is **das Papier,** and the pronoun used for it is **es** (*it*).

All three words—**er, sie** and **es**—will, when referring to things, be translated by English *it.*

Luckily the names of males of living things are normally masculine, and females, feminine.

The gender of a German noun is shown by the form of the definite article (*the*) used with it.

Thus masculine nouns are preceded by **der,**

feminine nouns are preceded by **die,**

and neuter nouns are preceded by **das.**

All nouns in the plural are preceded by **die.**

The pronoun referring to plural nouns (Eng. *they*) is **sie.**

When learning German nouns, one should always learn the appropriate article with each. Thus one should not know simply the word for *factory*—**Fabrik**—but that it is **DIE Fabrik.**

# 2

# DIE FABRIKARBEITER: ARTEN UND LÖHNE

Ein Arbeiter erhält jede Woche einen Lohn, eine Summe Geld, für seine Arbeit. Ein Arbeiter erhält mehr Geld, wenn er schneller arbeitet und mehr Artikel macht als gewöhnlich. Sein Lohn hängt von seiner Leistung ab.

Ein Mechaniker und ein Techniker bekommen einen Lohn, der wöchentlich gleich ist. Er hängt nicht von ihrer Leistung ab.

Ein Ingenieur empfängt sein Geld monatlich. Seine Bezahlung ist im allgemeinen höher als die der Handwerker.

Bei Krankheit erhalten alle durch einige Wochen hindurch ihren vollen Lohn.

## VOCABULARY

| | | | |
|---|---|---|---|
| **die Art, Arten** | the kind, sort, kinds | **erhält** | receives (from **erhalten,** to receive) |
| **der Lohn, Löhne** | payment, wages, payments | **HÄNGT von seiner Leistung AB** | *depends* on his output |
| **die Woche, Wochen** | the week, weeks | | |
| **die Summe** | the sum | **bekommen** | get, receive |
| **das Geld** | the money | **empfängt** | receives (from **empfangen,** to receive) |
| **die Arbeit** | the work | | |
| **der Artikel** | the article | | |
| **die Leistung** | performance, output | | |
| | | **jede** | each, every |
| **der Techniker** | the technician (*see Grammar, p.* 4) | **sein, seiner** | his |
| | | **ihrer, ihren** | their |
| **der Ingenieur** | the engineer (*see Grammar, p.* 4) | **mehr** | more |
| | | **wenn** | if, when, whenever |
| **die Bezahlung** | payment, remuneration | | |
| | | **schneller . . . als gewöhnlich** | more quickly than usual |
| **der Handwerker** | the manual worker | | |
| **die Krankheit** | illness | | |

3

| von | from (*usually*) but here it must be translated by on | höher als DIE der Hand- werker | higher than that of the manual workers |
| wöchentlich | weekly | bei Krankheit | in the event of ill- |
| monatlich | monthly | | ness |
| gleich | equal to, like, the same | vollen | full |

## Expressions to be learnt

| von etwas abhängen | to depend on something |
| im allgemeinen | in general |
| durch einige Wochen hindurch | for some weeks |

## Distinguish the workers in industry

| der Mechaniker | unqualified "engineer", mechanic |
| der Techniker | college trained and qualified engin-eer, technician |
| der Ingenieur | engineer with diploma obtained at university |

## GRAMMAR

### "EIN" WORDS—ACCUSATIVE CASE—VERB

**Ein** is the indefinite article.

It becomes **einen** when the *masculine* noun it precedes is a direct object, e.g. **Der Arbeiter bekommt EINEN Lohn.**

Before feminine nouns the form is **eine** when the noun is either subject or object.

Before neuter nouns **ein** is used before subject and object.

Words having the same case endings as **ein** are **kein** (*no*); **mein** (*my*); **sein** (*his, its*).

The case of the subject is called the *nominative*.

The case of the direct object is called the *accusative*.

| | M | F | N | Plural M, F, N. |
|---|---|---|---|---|
| N | ein | eine | ein | keine, meine |
| A | einen | eine | ein | keine, meine |

## The German Verb

The Infinitive of a German verb always ends in **-n,** usually in **-en.**

e.g. **sein**      *to be*          **machen**      *to make, do*
     **tun**       *to do*          **bekommen**  *to receive*

The form the verb has depends upon its subject and its tense.

**Der Arbeiter  MACHT**
*The worker*    *makes*
**Die Männer  MACHEN**
*The men*       *make*
**Der Arbeiter  MACHTE**  (*made*)
**Die Männer  MACHTEN** (*made*)

## The Present Tense of Verbs

The two commonest verbs are—

**sein**          *to be*
**haben**         *to have*

Present tense of **sein** and **haben**—

### SEIN

**ich bin**          *I am*
**er, sie, es ist**  *he, she, it is*
**wir sind**         *we are*
**Sie sind**         *you are*
**sie sind**         *they are*

### HABEN

**ich habe**          *I have*
**er, sie, es hat**   *he, she, it has*
**wir haben**         *we have*
**Sie haben**         *you have*
**sie haben**         *they have*

Regular verbs have endings like **machen** (*make*)—

| | |
|---|---|
| **ich mache** | *I make* |
| **er, sie, es macht** | *he, she, it makes* |
| **wir machen** | *we make* |
| **Sie machen** | *you make* |
| **sie machen** | *they make* |

But many verbs, especially commonly used verbs, are irregular. These must be noted and learnt individually.

Note the irregularity in

### ERHALTEN

| | |
|---|---|
| **ich erhalte** | *I receive* |
| **er, sie, es ERHÄLT** | *he, she, it receives* |
| **wir erhalten** | *we receive* |
| **Sie erhalten** | *you receive* |
| **sie erhalten** | *they receive* |

# 3

# WERKZEUGE

Der Mechaniker benutzt viele Werkzeuge. In seiner Werkstätte braucht er Hämmer, Meißel, Bohrer, Sägen, Schraubenzieher, Schraubenschlüssel, Schraubstöcke, Keile und so weiter.
Er braucht auch Instrumente zum Messen, z.B. Lineale und Schraublehren (Mikrometer).
Der Hammer ist ein Schlagwerkzeug. Mit ihm hämmert der Arbeiter das Metall. Mit dem Meißel bearbeitet er das Holz und das Metall. Er durchbohrt sie mit dem Bohrer. Mit der Säge schneidet er sie entzwei. Er schlägt den Meißel mit dem Hammer. Mit dem Schraubenzieher dreht er die Schrauben und mittels des Schraubenschlüssels löst er die Muttern oder zieht sie fest. Um das Werkstück zu bearbeiten hält er es im Schraubstock fest. Er mißt es mit dem Lineal oder mit der Schraublehre.

## VOCABULARY

| | | | | |
|---|---|---|---|---|
| das Werkzeug | tool | | das Messen | measuring, |
| der Hammer | hammer | | | measurement |
| der Meißel | chisel | | (*from verb* **messen** to measure. *Any* | |
| der Bohrer | borer, gimlet, | | *infinitive in German may be used as a* | |
| | | drill | *noun, e.g.* | |
| die Säge | saw | | das Schreiben | writing |
| der Schrauben- | screwdriver | | das Machen | making |
| zieher | | | das Bohren | boring |
| der Schrauben- | spanner | | | |
| schlüssel | | | *These infinitives used as nouns are* | |
| der Schraub- | vice | | *always neuter in gender)* | |
| stock | | | das Lineal | rule |
| der Keil | wedge | | das Schlagwerk- | striking-tool |
| das Instrument | instrument | | zeug | |
| die Schraub- | micrometer | | die Mutter | nut |
| lehre | | | das Werkstück | workpiece, job |

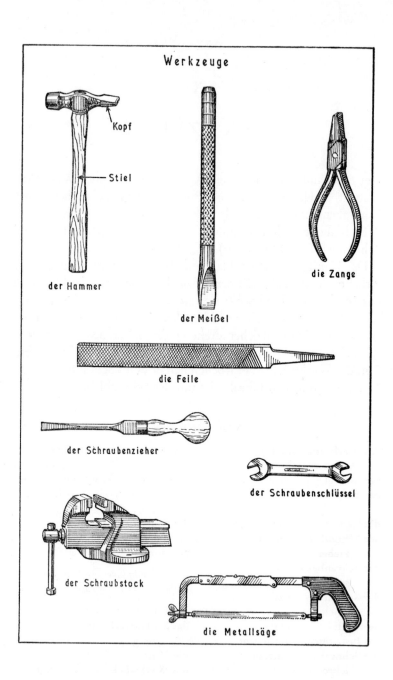

Werkzeuge

Kopf

Stiel

der Hammer

der Meißel

die Zange

die Feile

der Schraubenzieher

der Schraubenschlüssel

der Schraubstock

die Metallsäge

| | | | |
|---|---|---|---|
| **benutzen** | make use of, employ | **messen (mißt** is 3rd person singular) | measure |
| **brauchen** | need | | |
| **hämmern** | hammer | **und so weiter** | et cetera, etc. (abbreviated usually to **u.s.w.**) |
| **bearbeiten** | work, fashion | | |
| **durchbohren** | pierce | | |
| **schneiden** | cut | **z.B.** (short for **zum Beispiel**) | for example (e.g.) |
| **schlagen** | hit, strike | | |
| **drehen** | turn | **entzwei** | in two |
| **lösen** | loosen, unscrew, slacken | **mittels** | by means of (preposition followed by genitive case) |
| **ziehen (fest)** | tighten | | |
| **halten** | hold | **um . . . zu** | in order to |

## GRAMMAR

German nouns form their plurals in several ways. Though there is some regularity observable in this, it is perhaps best to learn the plural of each noun when learning its singular.

(*a*) Nouns which in the singular end in **-el, -en** and **-er** receive no ending to form the plural, but the internal vowel may be modified, e.g.

**der Meißel, die Meißel; der Hammer, die Hämmer**

but note—

**die Mutter** (*the nut*), **die Muttern**

(*b*) Nouns (usually *m.* and *n.*) sometimes form their plurals by adding **-e**, e.g.

**der Ingenieur, die Ingenieure; der Keil, die Keile;**
**das Werkzeug, die Werkzeuge; der Zahn, die Zähne**

(*c*) Neuter nouns often form their plurals by the addition of **-er**

**das Rad** (*wheel*), **die Räder**

(*d*) Feminine nouns often form their plurals by the addition of **-n** or **-en**

**die Säge, die Sägen; die Schraube, die Schrauben**

All German nouns end in **-n** in the dative plural.

### The Case Forms of the Definite Article

As we have seen, the subject (or nominative) case forms of the German *the* are—

|  | *Singular* |  | *Plural* |
| :---: | :---: | :---: | :---: |
| M | F | N | *all genders* |
| **der** | **die** | **das** | **die** |

The corresponding object (or accusative) case forms are—

| **den** | **die** | **das** | **die** |
| :---: | :---: | :---: | :---: |

The possessive or genitive case forms are—

| **des** | **der** | **des** | **der** |
| :---: | :---: | :---: | :---: |

These forms may usually be translated by *of the*.

The forms for the *dative case*, which follows many prepositions (e.g. **mit, von, nach**) are—

| **dem** | **der** | **dem** | **den** |
| :---: | :---: | :---: | :---: |

The dative is also the case of the indirect object—

**Hans gibt DEM Werkmeister die Muttern.** (*Hans gives the nuts TO THE foreman.*)

The case of a German noun is shown by the article which precedes it.

# 4

# IN DER WERKSTÄTTE

Die Werkstätte ist ein großes Gebäude. Sie ist sehr hell. Das Dach ist größtenteils aus Glas. Die Sonne scheint durch. Die Arbeiter können ihre Arbeit gut sehen. Am Tage ist die Werkstätte vom Sonnenlicht beleuchtet, aber in der Nacht und oft im Winter wird sie elektrisch beleuchtet. In der Werkstätte sind viele große Maschinen: Drehbänke, Bohrmaschinen, Schleifbänke u.s.w.

Ein Arbeiter steht vor jeder Maschine. Der Lärm der vielen Motoren und Maschinen ist so groß, daß man kaum die menschliche Stimme hört. Die Arbeiter sprechen wenig, und wenn sie sprechen, müssen sie den Lärm überschreien.

Ich gehe zu einem Arbeiter. Er steht vor einer Drehbank. Ich schreie in sein Ohr. „Was machen Sie da?" „Ich mache Räder aus diesem Zylinder aus Stahl. Während sich der Zylinder dreht, schneide ich mittels dieses Werkzeuges Scheiben ab, die später Riemenscheiben werden."

## VOCABULARY

| | | | |
|---|---|---|---|
| das Gebäude | building | die Stimme | voice |
| das Dach | roof | das Ohr | ear |
| das Glas | glass | das Rad, die | wheel, wheels |
| die Sonne | sun | Räder | |
| der Tag | day | der Zylinder | cylinder |
| das Sonnenlicht | sunlight | der Stahl | steel |
| die Nacht | night | die Scheibe | disc |
| der Winter | winter | die Riemen- | belt-pulley, |
| die Drehbank | lathe | scheibe | driving-wheel |
| die Bohr- | drilling machine | | |
| maschine | | scheinen | shine |
| die Schleifbank | grinding machine | können | be able, can |
| der Motor | engine, motor | sehen | see |
| | | beleuchten | light, illuminate |

11

# Werkzeuge

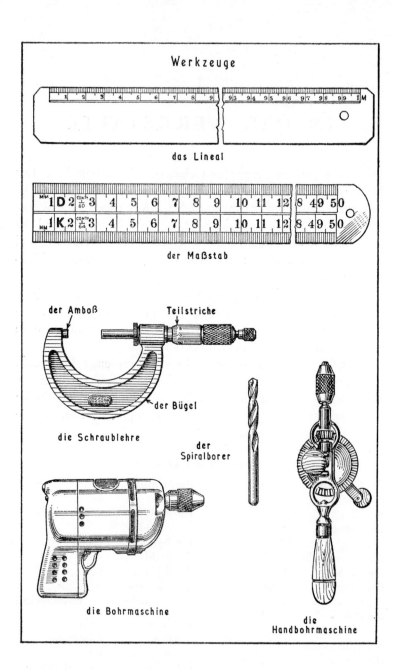

das Lineal

der Maßstab

der Amboß

Teilstriche

der Bügel

die Schraublehre

der
Spiralborer

die Bohrmaschine

die
Handbohrmaschine

| | | | |
|---|---|---|---|
| abschneiden, ich schneide ab | cut off (see Separable Verbs, Lesson 6) | durch | through |
| | | gut | good, well |
| | | am | contraction of an dem, in the, by |
| sich drehen | turn (see Reflexive Verbs, Lesson 12) | vom | contraction of von dem, by the |
| werden (wird) | (here) be (is) (see Grammar, p. 15) | im | contraction of in dem, in the |
| stehen | stand | | |
| hören | hear | elektrisch | electrically |
| müssen | have to, be obliged to, must | vor | in front of, before |
| | | man | one |
| überschreien | shout above | kaum | scarcely |
| gehen | go | menschlich | human |
| | | wenig | little |
| hell | light, bright | da | there |
| größtenteils | mainly | später | later |
| aus | of, out of | | |

## GRAMMAR

### ADJECTIVE USAGE

The adjective which follows the noun it qualifies—the "predicative" adjective—as here—

**Das Gebäude ist GROß**

is invariable in form.

The adjective which precedes the noun it qualifies agrees with it in gender and number, e.g.

| M | **gutER Erfolg** | *good success* |
|---|---|---|
| F | **großE Geschwindigkeit** | *great speed* |
| N | **hartES Metall** | *hard metal* |

M, F, N plural—**elektrischE Motoren,** *electric motors.*

If the adjective is preceded by the *indefinite article* **ein,** the form of the adjective is as above, i.e.

**ein gutER Erfolg, eine großE Geschwindigkeit, ein hartES Metall.**

Like forms follow **kein** (*no*) and possessive adjectives, e.g. **mein, sein.**

If, however, the adjective is preceded by the *definite article* the forms are as follows—

    M    **der gutE Erfolg**
    F     **die großE Geschwindigkeit**
    N    **das hartE Metall**

plural

    M    **die gutEN Erfolge**
    F     **die großEN Geschwindigkeiten**
    N    **die hartEN Metalle**

The **der** declension is used also with: **dieser, jeder, jener, welcher.**

The endings of the adjective in all cases, after **ein** and **der** words are shown in the table below.

### EIN + ADJECTIVE + NOUN

| | Masc. | Fem. |
|---|---|---|
| N | **ein großer Hammer** | **eine große Maschine** |
| A | **einen großen Hammer** | **eine große Maschine** |
| G | **eines großen Hammers** | **einer großen Maschine** |
| D | **einem großen Hammer** | **einer großen Maschine** |

| | Neuter | Plural M, F, N. |
|---|---|---|
| N | **ein hartes Metall** | **große Hämmer, Maschinen** |
| A | **ein hartes Metall** | **große Hämmer, Maschinen** |
| G | **eines harten Metalls** | **großer Hämmer, Maschinen** |
| D | **einem harten Metall** | **großen Hämmern, Maschinen** |

### DER + ADJECTIVE + NOUN

| | Masc. | Fem. |
|---|---|---|
| N | **der große Hammer** | **die große Maschine** |
| A | **den großen Hammer** | **die große Maschine** |
| G | **des großen Hammers** | **der großen Maschine** |
| D | **dem großen Hammer** | **der großen Maschine** |

| | Neuter | Plural M, F, N. |
|---|---|---|
| N | das harte Metall | die harten Hämmer, Metalle, etc. |
| A | das harte Metall | die harten Hämmer, Metalle |
| G | des harten Metalls | der harten Hämmer, Metalle |
| D | dem harten Metall | den harten Hämmern, Metallen |

### ADVERB

In German the adverb has the same form as the uninflected adjective, e.g.

**Der Junge arbeitet GUT,** *The boy works WELL.*
**Die Werkstätte wird ELEKTRISCH beleuchtet,** *The workshop is ELECTRICALLY lighted.*

### WERDEN

**Werden** is a very important verb.

(*a*) Its initial meaning is *become*, e.g.

**Die Tage werden länger,** *The days become longer.*

(*b*) It is the auxiliary verb used to form the Future Tense (see Lesson 7), e.g.

**Die Arbeiter werden essen,** *The workmen will eat.*

(*c*) It is the auxiliary verb used to form the Passive Voice (see Lesson 9), e.g.

**Das Metall wird gebohrt,** *The metal is drilled.*

# 5

## DIE FABRIK VON GESTERN

Vor fünfzig Jahren stand an der Straße eines Birminghamer Vorortes ein langes schmutziges Gebäude. Rund herum gab es zahllose primitive Häuser, die mit armen Leuten überfüllt waren. Im Sommer sah man die Kinder barfuß auf der Straße laufen.

Jeden Tag um 7 Uhr hörte man die Dampfpfeife der Fabrik, die die Arbeiter warnte, daß man die Tore zumache. Die Arbeiter, die später ankamen, wurden ausgeschlossen.

Bald sah man eine schwarze Rauchwolke über dem hohen Schlot hängen, und von der Fabrik kam ein großer Lärm. Es war das Hämmern, das Brausen und das Stampfen der Maschinen.

Die Bediener der Maschinen, die Arbeiter, wenn sie um 6 Uhr abends aus den Toren der Fabrik strömten, sahen blaß und müde aus. Ihre Gesichter, Hände und Kleider waren schmutzig. Sie mußten jeden Tag zehn Stunden arbeiten. Am Samstag hatten sie das Glück, nur sechs Stunden arbeiten zu müssen.

Hinter der Fabrik lagen große Haufen Kohle, der Brennstoff für die Dampfmaschinen, die der Fabrik die bewegende Kraft lieferten. Daneben häufte man die Asche der ausgebrannten Kohlen auf. Unter einem grauen Himmel war der Anblick trostlos.

Durch offene Türen sahen die Kinder Männer vor den großen Dampfkesseln stehen. Ihre halbnackten Körper glühten rot, wie sie die Kohlen auf die gierigen Flammen schaufelten.

### VOCABULARY

| | | | |
|---|---|---|---|
| **das Jahr** | year | **der Vorort** | suburb |
| **die Straße** | street, road | **die Leute** | people |
| **das Haus, die Häuser** | house, houses | **der Sommer** | summer |

16

| | | | |
|---|---|---|---|
| das Kind, die Kinder | child, children | ankommen | arrive |
| die Uhr | clock (um 7 Uhr at 7 o'clock) | hängen | hang |
| | | kommen | come |
| die Dampfpfeife | steam siren | strömen | stream |
| das Tor | gate | liegen | lie |
| die Rauchwolke | cloud of smoke | liefern | supply |
| der Schlot, die Schlöte | chimney, chimneys | häufen | heap (up) |
| | | glühen | glow |
| das Brausen | roar | schaufeln | shovel |
| das Stampfen | stamping, pounding | fünfzig | fifty |
| | | lang | long |
| der Bediener | servant | schmutzig | dirty |
| das Gesicht, die Gesichter | face, faces | zahllos | innumerable |
| | | primitiv | simple |
| die Hand, die Hände | hand, hands | arm | poor |
| | | überfüllt | over-filled |
| das Kleid, die Kleider | dress, clothes | barfuß | barefoot |
| | | daß | that (conjunction) |
| die Stunde | hour | spät | late |
| der Samstag | Saturday | ausgeschlossen | shut out |
| das Glück | luck, happiness | bald | soon |
| der Haufen | heap | schwarz | black |
| die Kohle | coal | über | over, above, about |
| der Brennstoff | fuel | hoch, hoh . . . | high |
| die Dampf-maschine | steam engine | abends | in the evening |
| | | blaß | pale |
| die Kraft | power | müde | tired |
| die Asche | ashes | hinter | behind |
| der Himmel | sky | bewegend | motive, moving |
| der Anblick | view | daneben | nearby |
| die Tür | door | ausgebrannt | burnt out |
| der Dampfkessel | steam boiler | unter | under |
| der Körper | body | grau | grey |
| die Flamme | flame | trostlos | cheerless, bleak |
| | | offen | open |
| laufen | run | halbnackt | half-naked |
| hören | hear | rot | red |
| warnen | warn | wie | as, how, like |
| zumachen | shut | gierig | greedy |

*Special points of vocabulary*

**rund herum**                              round about

**es gab** *Past tense of* **es gibt**      there was, there were.

**man sah die Kinder LAUFEN**    one saw the children *running*

In German the infinitive is used when we in English use the present participle. Note also in this passage—

**man sah eine Rauchwolke HÄNGEN,** one saw a cloud of smoke *hanging*.

**die Kinder sahen Männer STEHEN,** the children saw men *standing*.

**DIE die Arbeiter warnte,** *which* warned the workers.
**DIE später ankamen,** *who* came later.

In these two expressions **die** is a relative pronoun. For relative pronouns see p. 36.

> **wurden ausgeschlossen**      were shut out
> **das Hämmern**      the hammering
> **sahen blaß aus**      looked pale
> **aussehen** *is a separable verb (see* p. 23) *meaning* to appear
> **zu müssen**      to have to

## GRAMMAR

### SIMPLE PAST TENSE OF VERBS

*Regular* or "weak" verbs form their past tense by having a *t* attached to their stems before the endings determined by the subject, e.g. **machen.** Stem **mach.**

| *Present tense* | *Past tense* |
|---|---|
| **ich mache** (*make*) | **ich machTE** (*made*) |
| **er, sie, es macht** | **er, sie, es machTE** |
| **wir machen** | **wir machTEN** |
| **Sie machen** | **Sie machTEN** |
| **sie machen** | **sie machTEN** |

Note that in the past tense the form after **ich, er, sie** and **es** is the same.

*Irregular* verbs do not follow this rule, and the past tense of each has to be learned.

Simple past tense of some common irregular verbs—

| **SEIN** | **WERDEN** | **STEHEN** | **SEHEN** |
|---|---|---|---|
| ich war | ich wurde | ich stand | ich sah |
| er war | er wurde | er stand | er sah |
| wir waren | wir wurden | wir standen | wir sahen |
| Sie waren | Sie wurden | Sie standen | Sie sahen |
| sie waren | sie wurden | sie standen | sie sahen |

Note again that there are only two forms of the past tense—singular and plural. The form **du**—*thou*—is not used in this book, though **du** is by no means obsolescent in German. It is not likely, however, to occur in technical literature.

# 6

## DIE FABRIK VON HEUTE

Heute steht in einem schönen Vorort von Birmingham ein sauberes, helles vierstöckiges Gebäude an einer breiten Straße, die mit Bäumen bepflanzt ist.

Man sieht keinen hohen Schlot, keinen Rauch. Um acht Uhr früh strömen die Arbeiter zu den breiten Toren der Fabrik. Diese Menschen sehen sauber und gesund aus. Viele kommen in ihren eigenen Wagen an, die sie auf einem großen der Fabrik gehörenden Parkplatz abstellen. Sie treten arbeitswillig in die Fabrik ein.

Kurz nach acht Uhr hört man das Summen der elektrischen Motoren, die die Maschinen antreiben.

Um neun Uhr kommt das Büropersonal der Fabrik an. Es sind Schreiber, Ingenieure und Zeichner.

Die Handwerker und Mechaniker arbeiten täglich von acht Uhr früh bis fünf Uhr abends an fünf Tagen der Woche. Die Mittagspause dauert eine Stunde. Das Büropersonal arbeitet von neun bis fünf Uhr, das heißt, wöchentlich fünf Stunden weniger als die anderen.

Wenn sie die Fabrik betreten, und wenn sie sie abends verlassen, sind alle Arbeiter sauber und nett gekleidet. Innerhalb der Fabrik tragen die Handwerker schonende Arbeitsanzüge. Wenn sie durch die Arbeit schmutzig werden, können sie sich leicht waschen, weil es genügend Waschräume mit heißem Wasser gibt.

Mittags hört die Arbeit auf und alle gehen entweder nach Hause oder in die schönen Speisesäle, um dort zu essen. Im Speisesaal kann man um einen niedrigen Preis eine schmackhafte Mahlzeit einnehmen.

Nach der Tagesarbeit sind die Arbeiter noch frisch, und wenn sie wollen, können sie noch Sport auf den Sportplätzen der Fabrik treiben. Einige gehören den verschiedenen Klubs an, die sich für Theaterstücke, Opern, die Photographie oder das Wandern interessieren. Die Firma unterstützt diese kulturellen Interessen ihrer Arbeiter mit Geld.

## VOCABULARY

| | | | |
|---|---|---|---|
| der Baum, Bäume | tree, trees | die Firma, Firmen | firm, firms |
| der Rauch | smoke | das Interesse, die Interessen | interest(s) |
| der Mensch, Menschen | human being, people | | |
| der Wagen | car | bepflanzen | plant; **bepflanzt,** planted |
| der Parkplatz | car-park | | |
| das Summen | hum, humming | aussehen (Sep. verb) | look, appear |
| das Büro- personal | office staff | abstellen (Sep. verb) | put, put away |
| der Schreiber | clerk | antreiben (Sep. verb) | drive, motivate |
| der Zeichner | draughtsman | | |
| die Mittags- pause | midday break | dauern | last |
| der Arbeits- anzug (-züge) | overall(s) | betreten | enter |
| | | verlassen | leave |
| der Waschraum, Waschräume | washroom, wash- rooms | kleiden | dress; **gekleidet,** dressed |
| das Wasser | water | tragen | wear, carry |
| der Speisesaal, Speisesäle | dining room, dining rooms | waschen | wash |
| | | essen | eat |
| der Preis | price | einnehmen | take, receive |
| die Mahlzeit, Mahlzeiten | meal, meals | wollen | be willing, want |
| | | treiben | drive |
| der Sport (-e) | sport | *but note the idiom:* **einen Sport treiben** to engage in a sport | |
| der Sportplatz, Sportplätze | sportsground(s) | | |
| der Klub (-s) | club | unterstützen | support |
| das Theater- stück (e) | play, drama | heute | today |
| die Oper (-n) | opera | sauber | clean |
| die Photo- graphie | photography | hell | bright, light |
| | | vierstöckig | four-storey |
| das Wandern | rambling | breit | broad |
| | | gesund | healthy |
| | | eigen | own |

| | | | |
|---|---|---|---|
| **arbeitswillig** | ready and willing to work | **heiß** | hot |
| **kurz** | short, shortly | **entweder . . . oder** | either . . . or |
| **nach** | after | **schön** | fine, beautiful |
| **täglich** | daily | **niedrig** | low |
| **weniger (als)** | less (than) | **schmackhaft** | tasty |
| **die anderen** | the others | **frisch** | fresh |
| **nett** | neat, nice (ly) | **noch** | still |
| **schonend** | protective | **einige** | some |
| **leicht** | easily | **verschieden** | different, various |
| **weil** | because | **kulturell** | cultural |
| **genügend** | enough, adequate | | |

## Special points of vocabulary

| | |
|---|---|
| **um acht Uhr** | at eight o'clock |
| **acht Uhr früh** | 8 a.m. |
| **das heißt,** often contracted to **d.h.** | that is, i.e. |
| **weniger als** | less than |
| **innerhalb** (*preposition followed by gen.*) | inside |
| **es gibt** | there are |
| **mittags** | at noon |
| **nach Hause** | home |
| **sich interessieren für** (reflexive verb) | be interested in |
| **ankommen** | arrive |
| **eintreten** | enter |
| **aufhören** | stop |
| **angehören**(*followed by dative*) | belong |
| **der Fabrik GEHÖREND** | *belonging* to the factory |

Separable verbs. See Grammar Section, p. 23

## GRAMMAR

### PREPOSITIONS

In English all prepositions "govern," i.e. are followed by, the accusative or object case, e.g. to *him*, by *her*, in *them*, but in German some prepositions govern the accusative case, others the dative case, others the genitive case and some, according to the circumstances, govern either the accusative or the dative case. To master this difficulty it is necessary to learn *by heart* each group of prepositions. Lists may be found at the end of this chapter.

## WORD ORDER

1. *Normal*

Subject followed by verb, e.g.

**Die Sonne scheint.**

2. *Inverted*

If the sentence begins with anything other than the subject, the verb is placed before the subject, e.g.

**Heute scheint die Sonne.**

A phrase or a clause has the same effect as a word, e.g.

**Bei schönem Wetter scheint die Sonne.**

3. *Transposed*

In subordinate clauses the finite verb is placed at the end, e.g.

**Wenn sie die Fabrik BETRETEN, sind alle Arbeiter rein.**

### SEPARABLE VERBS

Some verbs have a prefix which in simple tenses becomes detached and is placed at the end of the sentence, e.g.

**ankommen,** *to arrive.*
**Viele kommen in ihren eignen Wagen AN.**
**aussehen,** *to appear.*
**Die Arbeiter sahen rein AUS.**

### LISTS OF PREPOSITIONS

1. *Governing the Accusative Case*

| | |
|---|---|
| **durch** | through, by |
| **für** | for |
| **gegen** | against, towards |
| **ohne** | without |
| **um** | about, round, at |
| **wider** | against |

## 2. *Governing the Dative Case*

| | |
|---|---|
| **aus** | out of |
| **außer** | besides |
| **bei** | at, with, by |
| **mit** | with |
| **nach** | after, to, towards, according to |
| **seit** | since |
| **von** | from, by, of |
| **zu** | to |
| **entgegen*** | towards |
| **gegenüber*** | opposite |

\* Used after noun or pronoun, e.g.

**Hans kommt dem Werkführer entgegen.**
**Die Kirche steht der Fabrik gegenüber.**


## 3. *Governing the Genitive Case*

| | |
|---|---|
| **inmitten** | in the midst of |
| **anstatt, statt** | instead of |
| **außerhalb** | outside |
| **innerhalb** | inside |
| **längs** | along |
| **infolge** | in consequence of, as a result of |
| **diesseits** | on this side of |
| **jenseits** | on that side of, beyond |
| **hinsichtlich** | with reference to |
| **trotz** (sometimes also with dative) | in spite of |
| **wegen** | on account of, because of |
| **während** | during |
| **mittels** | by means of |
| **oberhalb** | above |
| **unterhalb** | below |
| **seitens** | on the part of |

4. *Governing sometimes the Accusative and sometimes the Dative*

| | |
|---|---|
| **an** | at, on, to |
| **auf** | on, in, to |
| **hinter** | behind |
| **in** | in, into |
| **neben** | near |
| **über** | over |
| **unter** | under |
| **vor** | before, in front of |
| **zwischen** | between |

N.B. If motion towards the object is indicated, the Accusative is used, e.g.

**Die Arbeiter gehen in DIE Fabrik.**

If no motion *towards* the object is implied, the Dative is used, e.g.

**Die Arbeiter arbeiten in DER Fabrik.**

# 7

# DIE ZUKUNFT DES INGENIEURWESENS

In der Vergangenheit, besonders zwischen dem 15. und dem 17. Jahrhundert, war die Entwicklung des Ingenieurwesens sehr langsam, aber seit der industriellen Umwälzung des späten 18. Jahrhunderts ist das Entwicklungstempo immer schneller geworden.

In der Gegenwart werden Maschinen und Vorrichtungen aller Art sehr schnell altmodisch.

Das Denken des Ingenieurs ist auf die Zukunft gerichtet. Alles muß billiger, schneller und leistungsfähiger erzeugt werden.

Was können wir erwarten, wenn wir unseren Blick auf die nächste Zukunft werfen?

Man wird viele Automaten erfinden, um Dinge schneller und billiger zu erzeugen.

Viele Arbeiter werden dadurch überflüssig werden. Man macht schon Rechenmaschinen, die komplizierte mathematische Probleme schneller und akkurater als die Menschen lösen.

Der Metallurge wird immer stärkere Metalle erzeugen, damit die daraushergestellten Fahrzeuge und Flugzeuge schneller, länger und weiter fahren und fliegen können, und damit sie auch größere Spannungen unter ganz neuen Verhältnissen ertragen können.

Der Chemiker und der Physiker werden neue Brennstoffe, neue Energiequellen aussuchen, um die jetzigen Quellen zu ersetzen, die schnell erschöpft sein werden.

## VOCABULARY

| | | | |
|---|---|---|---|
| **die Zukunft** | the future, the "to-come" | **die Vergangen- heit** | the past |
| **das Ingenieur- wesen** | engineering | **das Jahrhundert** | century |
| | | **die Entwicklung** | development |

| | | | |
|---|---|---|---|
| die Umwälzung | revolution | fahren | travel |
| das Tempo | speed | fliegen | fly |
| die Gegenwart | present | ertragen | bear, tolerate |
| die Vorrichtung | device | aussuchen | seek out |
| das Denken | thinking | ersetzen | replace |
| der Blick | glance | erschöpfen | exhaust |
| der Automat | automatic mach-ine, automation | besonders | especially |
| | | langsam | slow(ly) |
| das Ding | thing | immer | always, ever, more and more |
| die Rechen-maschine | calculating mach-ine, computer | | |
| das Problem | problem | alles | everything |
| der Metallurge | metallurgist | altmodisch | old-fashioned |
| das Fahrzeug | vehicle | billig | cheap |
| das Flugzeug | aeroplane | leistungsfähig | efficient, produc-tive |
| die Spannung | tension | | |
| das Verhältnis | condition, circum-stance | was | what |
| | | unser | our |
| der Chemiker | chemist | nächst | next, nearest |
| der Physiker | physicist | (superlative of adjective nah, near) | |
| die Energie-quelle | source of energy | | |
| die Quelle | source | dadurch | through it, "therethrough" |
| richten (p.p. gerichtet) | direct | überflüssig | superfluous |
| | | schon | already |
| erzeugen | produce | kompliziert | complicated |
| erwarten | expect | mathematisch | mathematical |
| werfen | cast, throw | akkurat | accurate |
| erfinden | invent | ganz | quite, whole |
| lösen | solve | neu | new |
| herstellen (p.p. hergestellt) | produce, manu-facture | jetzig | present, existing |

*For special note*

| | |
|---|---|
| immer schneller | faster and faster, more and more rapid |
| dadurch | through this |
| damit (conjunction here) | so that, in order that |
| daraushergestellt | manufactured from it |
| um zu ersetzen | in order to replace |

## GRAMMAR

### The Future Tense

In English we compose the Future Tense using the auxiliaries *shall* and *will* followed by the verb, thus—

*I shall go*

*He will go*

In German the Future Tense is composed by using the auxiliary verb **werden** with the infinitive form of the verb, thus—

| | |
|---|---|
| ich **WERDE** gehen | *I shall go* |
| er, sie, es, **WIRD** gehen | *he, she, it will go* |
| wir **WERDEN** gehen | *we shall go* |
| Sie **WERDEN** gehen | *you will go* |
| sie **WERDEN** gehen | *they will go* |

*Auxiliary verbs of mood*

The following verbs, called "auxiliary verbs of mood" or "modal auxiliaries," are similarly followed by an infinitive verb —as their counterparts in English usually are—

| | |
|---|---|
| **KÖNNEN** | *can, to be able to* |
| **DÜRFEN** | *may, to be permitted to* |
| **SOLLEN** | *shall, ought to, to be expected to* |
| **MÜSSEN** | *must, to have to* |
| **WOLLEN** | *will, wish to, want to* |
| **MÖGEN** | *like, like to, care to* |

*Distinguish between—*

| | |
|---|---|
| Er **WIRD** kommen | *He will (certainly) come* |
| and Er **WILL** kommen | *He wishes to, intends to, come* |

N.B. In a simple sentence in the future the infinitive of the verb is placed at the end of the sentence, e.g.

**Der Ingenieur wird sicher eine Lösung FINDEN,** *The engineer will certainly find a solution*

## Comparison of Adjectives and Adverbs

When comparing things with like qualities in differing degrees we often use special forms of the adjective. Of two *green* objects we say one is *greener* than the other. If we have three or more green objects the one that has the highest degree of greenness is said to be the *greenest*.

The form *green* is said to be the *positive* form.

The form *greener* is called the *comparative* form.

The form *greenest* is called the *superlative* form.

German has similar forms, some of which, as in English, are irregular.

*Some examples*

| Positive | Comparative | Superlative |
|----------|-------------|-------------|
| **grün** | **grüner** | **grünst** |
| **schnell** | **schneller** | **schnellst** |
| **billig** | **billiger** | **billigst** |
| **hoch** | **höher** | **höchst** |
| **lang** | **länger** | **längst** |
| **nah** | **näher** | **nächst** |
| **groß** | **größer** | **größt** |
| **gut** | **besser** | **best** |

*Note the following constructions—*

**immer größer**                    *bigger and bigger*

**Der Schlot ist HÖHER ALS die Fabrik,** *The chimney is HIGHER THAN the factory*

**Das Haus ist NICHT SO GROSS wie die Fabrik,** *The house is NOT SO LARGE AS the factory*

*Adverbs* are compared in the same way as adjectives, except in the Superlative, when the forms are: **am grünsten, am schnellsten, am besten,** etc., e.g.

**Diese Maschine erzeugt AM SCHNELLSTEN,** *This machine produces fastest*

# 8

## WIE EIN DEUTSCHER INGENIEUR WIRD

Ein Junge, der Ingenieur werden will, bleibt in der Schule—dem Realgymnasium—bis er 18 Jahre alt ist. Er muß gute Noten aus Mathematik, aus Physik und aus Deutsch bekommen, wenn er die Reifeprüfung macht. Er geht dann auf die Technische Hochschule—eine Art technischer Universität—wo er mindestens vier Jahre lang studiert. Am Ende der Studienzeit erhält der junge Mann sein Diplom.

Als Student bereitet er sich für seinen zukünftigen Beruf nicht nur durch das Studium der Mathematik und der Naturwissenschaften sondern auch der neueren Sprachen vor. Er muß auch praktische Übungen in der Werkstätte machen, so daß er die Probleme der Handarbeiter versteht, die er später vielleicht wird dirigieren müssen.

### VOCABULARY

| | | | |
|---|---|---|---|
| **der Junge** | boy | **die Natur-** | science |
| **die Schule** | school | **wissenschaft** | |
| **die Note** | mark | **die Sprache** | language |
| **die Mathematik** | mathematics | **die Übung** | exercise |
| **die Physik** | physics | **der Hand-** | manual worker |
| **das Deutsch** | German | **arbeiter** | |
| **die Universität** | university | | |
| **das Ende** | end | **bleiben** | remain, stay |
| **die Studienzeit** | period of study | **studieren** | study |
| **das Diplom** | diploma, degree | **verstehen** | understand |
| **der Student, die** | student, students | **dirigieren** | direct |
| **Studenten** | | | |
| **der Beruf** | profession | **bis** | until |
| **das Studium,** | study, studies | **alt** | old |
| **die Studien** | | **dann** | then |

| technisch | technical, technological | sondern | but (after a negative) |
| wo | where | neu (comp. neuer) | new, modern |
| mindestens | at least | | |
| jung | young | praktisch | practical |
| zukünftig | future | vielleicht | perhaps |
| nur | only | | |

*For special note*

| das Realgymnasium | secondary school teaching Latin, modern languages and science |
| die Reifeprüfung | a final school leaving examination, somewhat similar to G.C.E. |
| sich vorbereiten | to prepare (**vorbereiten** is separable). Lit: to prepare oneself (*see* Grammar, p. 51) |

## GRAMMAR

### PERSONAL PRONOUNS

Like their English equivalents, the German personal pronouns have object forms, e.g. **er** (*he*) has the object form **ihn** (*him*), but they have also special forms for the *indirect object*. In the sentence, *The foreman handed him a tool*, the direct object is *a tool*; *him* is the indirect object. In German it has the form **IHM.**

The indirect object is in the *dative case*.

The dative forms of the pronoun follow the prepositions which require a dative (*see* p. 24).

In this book the most commonly occurring pronouns are those of the third person (*he, she, it, they*).

*Third Person Pronouns*

| | Singular | | | | Plural | |
| | M | F | N | | M.F.N. | |
| Nom. | er | sie | es | (*he, she, it*) | sie | (*they*) |
| Acc. | ihn | sie | es | (*him, her, it*) | sie | (*them*) |
| Dat. | ihm | ihr | ihm | (*to him, to her, to it*) | ihnen | (*to them*) |
| Possessive | sein | ihr | sein | (*his, her, its*) | ihr | (*their*) |

### Other Personal Pronouns

| | | | | | | | |
|---|---|---|---|---|---|---|---|
| Nom. | **ich** | (*I*) | **wir** | (*we*) | **Sie** | (*you*) |
| Acc. | **mich** | (*me*) | **uns** | (*us*) | **Sie** | (*you*) |
| Dat. | **mir** | (*me, to me*) | **uns** | (*us, to us*) | **Ihnen** | (*to you*) |
| Posses-sive | **mein** | (*my*) | **unser** | (*our*) | **Ihr** | (*your*) |

## Possessive Adjectives

Each pronoun has a corresponding adjective, e.g. *I, my; he, his; they, their.* These are the possessive adjectives. The German possessive adjectives are as follows—

| | | | |
|---|---|---|---|
| **mein** | *my* | **unser** | *our* |
| **sein** | *his or its* | **Ihr** | *your* |
| **ihr** | *her or its* | **ihr** | *their* |

They are modified, "declined" like **ein** words, e.g.

N   **sein zukünftiger Beruf**
A   **seinen zukünftigen Beruf**
G   **seines zukünftigen Berufs**
D   **seinem zukünftigen Beruf**

(*see* Grammar of Chapter 4).

# 9

# DIE DREHBANK

Keine Maschine ist dem Mechaniker nötiger oder nützlicher als die Drehbank. Eine Werkstätte ohne diese vielseitige Werkzeugmaschine ist undenkbar. Obgleich sie so modern aussieht, hat die Drehbank eine lange Geschichte.

Die Drehbank ist in erster Linie eine Maschine zum Halten und Drehen eines Materials während es durch ein Werkzeug bearbeitet wird.

Die älteste Darstellung einer Drehbank findet sich auf einem ägyptischen Grabrelief aus der Zeit um 300 v.Chr. Bei dieser primitivsten Form, die noch heute in gewissen unterentwickelten Ländern vorkommt, wird um das Werkstück eine Schnur geschlungen, die mit der Hand oder mit einem Fiedelbogen hin- und her- bewegt wird.

Die durch ein Rad bewegte Drehbank ist erst im 14. Jahrhundert aufgekommen, und die Drehbank mit Zugspindeln zum Vortrieb des Werkzeugschlittens für das Schneiden von Gewinden wird dem erfinderischen Leonardo da Vinci zugeschrieben. Kunsttischler des 16. Jahrhunderts stellten damit hölzerne Zierstücke in Schraubenform her.

Erst im Jahre 1772 ließ man das Werkstück zentral ununterbrochen rotieren. Die Spindel wurde von einem Schwungrad mittels einer Schnur gedreht. Das Schwungrad wurde anfangs mit der Hand bewegt. Dadurch wurde eine konstante Bewegung in unverändertem Drehsinn erzielt.

Einem Engländer, John Wilkinson, verdanken wir die Erfindung der Präzisionsdrehbank. Er wollte die Zylinder ausbohren, die James Watt für seine Dampfmaschinen brauchte. Zylinder mit einem Durchmesser bis auf 127 Zentimeter wurden damit exakt ausgebohrt.

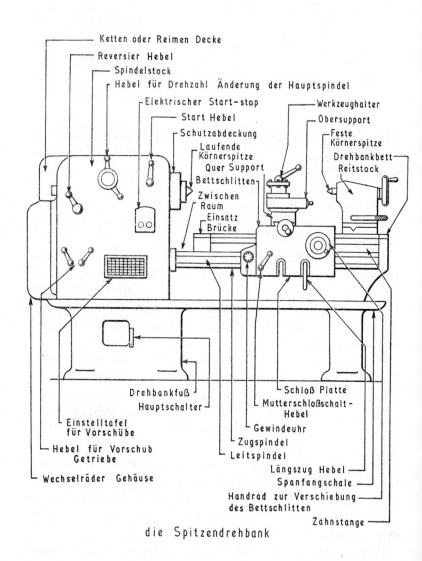

Ketten oder Reimen Decke

Reversier Hebel

Spindelstock

Hebel für Drehzahl Änderung der Hauptspindel

Elektrischer Start-stop

Start Hebel

Schutzabdeckung

Laufende
Körnerspitze
Quer Support

Bettschlitten

Zwischen
Raum

Einsatz
Brücke

Werkzeughalter

Obersupport

Feste
Körnerspitze

Drehbankbett

Reitstock

Drehbankfuß

Hauptschalter

Schloß Platte

Mutterschloßschalt-
Hebel

Einstelltafel
für Vorschübe

Gewindeuhr

Hebel für Vorschub
Getriebe

Zugspindel

Leitspindel

Wechselräder Gehäuse

Längszug Hebel

Spanfangschale

Handrad zur Verschiebung
des Bettschlitten

Zahnstange

die Spitzendrehbank

34

Ein anderer Engländer, Henry Maudslay, aus London, erfand im Jahre 1797 einen Werkzeugschlitten, der durch eine Zugspindel über das Bett der Drehbank bewegt wird.

## VOCABULARY

| | | | |
|---|---|---|---|
| die Drehbank | lathe | die Spindel | spindle |
| die Geschichte (-n) | story, history | das Schwungrad | fly-wheel |
| das Material (-ien) | material | die Bewegung | movement |
| | | der Drehsinn | direction of turning |
| die Darstellung (-en) | representation | der Engländer | Englishman |
| das Grabrelief | relief on a tomb | die Erfindung | invention |
| die Zeit (-en) | time | die Präzision | precision |
| die Form (-en) | form | der Zylinder | cylinder |
| das Land, Länder | land, lands; country | der Durchmesser | diameter |
| die Schnur (-en) | string or cord | das Bett (-en) | bed |
| der Fiedelbogen (-bögen) | fiddle-bow | sich finden | |
| das Rad, Räder | wheel | (*lit.*) to find itself, oneself, etc., to be found (*see* For *special note* below) | |
| die Zugspindel (-n) | feed screw | vorkommen | occur, happen |
| der Vortrieb (-e) | propulsion | schlingen, p.p. geschlungen | wind, sling |
| der Werkzeug- schlitten | tool slide | bewegen | move |
| das Schneiden | cutting | aufkommen aufgekommen | (p.p. rise, appear) |

(any verb can be made into a noun by prefixing **das** to the infinitive and spelling with a capital. Such nouns correspond to the English verbal nouns in -*ing*, e.g. **das Schreiben,** writing; **das Drehen,** turning)

| | |
|---|---|
| das Gewinde | thread |
| der Kunsttisch- ler | cabinet-maker |
| das Zierstück | decorative component |
| die Schrauben- form | screw-form |

| | |
|---|---|
| herstellen | produce, manufacture (*see* below) |
| lassen | let, cause to |

p.t. **ließ: man ließ das Werkstück rotieren** one caused the workpiece to rotate

| | |
|---|---|
| erzielen | achieve |
| ausbohren | bore |
| nötig | necessary |
| nützlich | useful |
| vielseitig | many-sided, versatile |
| undenkbar | unthinkable |
| obgleich | although |
| während | while |
| ägyptisch | Egyptian |

| | |
|---|---|
| **v.Chr. (vor Christi Geburt)** | B.C. |
| **dieser** (like **der**) | this |
| **primitiv** | very early |
| **gewiß** | certain |
| **unterentwickelt** | under-developed, backward |
| **hin und her** | backwards and forwards, to and fro |
| **erst** | first |
| **erfinderisch** | inventive |
| **hölzern** | wooden |
| **zentral** | centrally |
| **ununterbrochen** | uninterruptedly |
| **anfangs** | at first |
| **konstant** | constant |
| **unverändert** | unaltered |
| **exakt** | with exactitude |

*For special note*

| | |
|---|---|
| **in erster Linie** | primarily |
| **zum Halten** | to hold; *lit:* for the holding (of) |
| **findet sich** | is to be found; *lit:* finds itself |
| **noch heute** | even today |
| **die durch ein Rad BEWEGTE Drehbank** | the lathe *motivated* by a wheel |
| **einem etwas zuschreiben** | to ascribe something to somebody |
| **stellten . . . her** | produced, made; from **herstellen** (sep. verb) |
| **dadurch** | by this means, through this |
| **einem etwas verdanken** | to owe something to somebody; to be grateful to someone for something |

## GRAMMAR

### RELATIVE PRONOUNS

In English the common relative pronouns are *who* (*whom, whose*), *which* and *that*.

*The engineer WHO learns German is wise.*

*The turbine THAT Benoit Fourneyron invented was a water turbine.*

*The hammer WHICH you are looking for is here.*

*The flywheel WHOSE diameter is 25 cm regulates the speed.*

In German all these words may be translated by **der, die, das** in their various cases in singular and plural. The forms of this relative pronoun differ in the genitive and dative cases from those of the definite article.

|        | M      | F     | N      | Plural M.F.N. |
|--------|--------|-------|--------|---------------|
| Nom.   | der    | die   | das    | die           |
| Acc.   | den    | die   | das    | die           |
| Gen.   | dessen | deren | dessen | deren         |
| Dat.   | dem    | der   | dem    | denen         |

The following sentences illustrate their use—

**Der Ingenieur, DER Deutsch lernt, ist weis,** *The engineer who learns German is wise.*

**Die Turbine, DIE B. Fourneyron erfand, war eine Wasserturbine,** *The turbine that B. Fourneyron invented was a water turbine.*

**Der Hammer, DEN Sie suchen, ist hier.** *The hammer which you are looking for is here.*

**Das Schwungrad, DESSEN Durchmesser 25 Zentimeter ist, reguliert die Geschwindigkeit,** *The flywheel whose diameter is 25 cm regulates the speed.*

It will be observed that the relative pronoun agrees in gender and number with its antecedent (the noun to which it refers). Its case depends upon its function in the clause which it introduces.

### The Passive Voice of Verbs

In a sentence in which the subject *performs* the action, the verb is said to be *active*. In each of the following sentences the verbs in capitals are *active*.

> *The motor DRIVES the spindle* (Present tense)
> *The motor DROVE the spindle* (Simple past tense)
> *The motor HAS DRIVEN the spindle* (Perfect tense)

By making the objects of these sentences the subjects of some new ones, we can say what happens in another way. The subjects are now the recipients of the action. The forms of the verbs in capitals in the sentences below are said to be *passive*.

> *The spindle IS DRIVEN by the motor* (Present tense)
> *The spindle WAS DRIVEN by the motor* (Past tense)
> *The spindle HAS BEEN DRIVEN by the motor* (Perfect tense)

In English, therefore, the passive is formed by combining the appropriate number and tense forms of the verb "to be" with the *past participle* of the verb.

In German, the passive is formed by combining the appropriate forms of the verb **werden** with the *past participle* of the verb, thus—

**Die Spindel WIRD GETRIEBEN** (Present tense)
*The spindle is driven*
**Die Spindel WURDE GETRIEBEN** (Past tense)
*The spindle was driven*
**Die Spindel IST GETRIEBEN WORDEN** (Perfect tense)
*The spindle has been driven*

It is to be noted that the auxiliary verb, i.e. one of the various forms of **werden,** is often separated from the *past participle* by other words, because it is the rule that the past participle is placed at the end of the (simple) sentence, e.g.

**Die Spindel WIRD von dem Motor GETRIEBEN**
*The spindle IS DRIVEN by the motor*

# 10

# DIE KUNST DES ZEICHNERS

Wenn der Mechaniker in der Werkstätte ein Maschinenwerkzeug oder den Musterbestandteil eines neuen Artikels macht, beachtet er eine Zeichnung davon, die ihm alle notwendigen Einzelheiten der Form und der Dimensionen zeigt.

Diese Zeichnung kommt ursprünglich aus der Hand eines geschickten Zeichners, der im Konstruktionsbüro arbeitet.

Von dem Erfinder oder dem Konstrukteur erhält der Zeichner den Entwurf, eine Rohzeichnung des Werkzeuges oder des Bestandteiles, und davon entwickelt er sorgfältig maßstabgerechte Pläne und Projektionen, die der Arbeiter im Metall nachahmen kann.

Erst im Konstruktionsbüro werden also die Erfindungen lebendig—wenn man dort keine wesentlichen Schwächen in ihnen entdeckt.

Nachdem ein Meisterzeichner die erste Bleistiftzeichnung angefertigt hat, stellt man mit Tinte auf Papier oder durchsichtigem Tuch eine Pause her. Mittels dieser Pause macht der Bürophotograph Lichtpausen, so viel man will. Die Pausen werden oft von den jüngeren Zeichnern oder Zeichnerinnen angefertigt.

Das Werkzeug des Zeichners besteht aus dem Reißbrett, der Reißschiene, dem Dreieck, dem Winkelmesser, Linealen, Reißfedern und Zirkeln verschiedener Arten.

Während der junge Ingenieur einen Teil seiner Lehrzeit im Konstruktionsbüro verbringt, lernt er viel Nützliches, denn er sieht dort Pläne der verschiedensten Maschinen von ihren ersten Anfängen bis zur Vollendung.

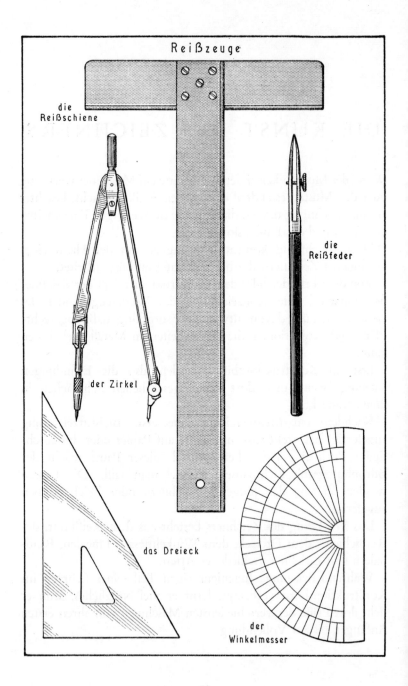

Reißzeuge

die
Reißschiene

der Zirkel

die
Reißfeder

das Dreieck

der
Winkelmesser

40

## VOCABULARY

| | | | |
|---|---|---|---|
| **die Kunst** | art, skill | **der Plan (Pläne)** | plan |
| **der Musterbe-** | prototype com- | **die Projektion** | projection |
| **standteil** | ponent | **die Schwäche** | weakness |
| **die Zeichnung** | drawing | **der Meister-** | expert draughts- |
| **die Einzelheit** | detail | **zeichner** | man |
| **das Konstruk-** | drawing office | **die Bleistift-** | pencil drawing |
| **tionsbüro** | | **zeichnung** | |
| **der Erfinder** | inventor | **die Tinte** | ink |
| **der Konstruk-** | designer | **das Papier** | paper |
| **teur** | | **das Tuch** | cloth |
| **der Entwurf** | sketch | **die Pause** | tracing |
| **die Rohzeich-** | rough drawing | | |
| **nung** | | | |

| | |
|---|---|
| **der Bürophotograph** | office photographer |
| **die Lichtpause** | photographic print |
| **das Reißbrett** | drawing board |
| **die Reißschiene** | tee-square |
| **das Dreieck** | set square |
| **der Winkelmesser** | protractor |
| **die Reißfeder** | drawing pen |
| **der Zirkel** | pair of compasses |
| **der Teil** | part |
| **die Lehrzeit** | apprenticeship |
| **der Anfang (Anfänge)** | beginning |
| **die Vollendung** | completion |

| | | | |
|---|---|---|---|
| **beachten** | pay attention to, study | **davon** | of it, from it |
| **zeigen** | show | **notwendig** | necessary |
| **entwickeln** | develop | **ursprünglich** | originally |
| **nachahmen** | imitate, copy | **geschickt** | clever |
| **entdecken** | discover | **sorgfältig** | carefully |
| **anfertigen** (sep.) | prepare | **maßstabgerecht** | scale, true to scale |
| **bestehen (aus)** | consist (of) | **also** | then, therefore |
| **+ dative** | | **lebendig** | living, alive |
| **verbringen** | spend (time) | **wesentlich** | essential |
| **lernen** | learn | **durchsichtig** | transparent |

*For special note*

| | | |
|---|---|---|
| **wenn** | when | } |
| **nachdem** | after | } |
| **während** | while | } |

"Subordinating" Conjunctions. See the Grammar Section which follows.

Nachdem ein Meisterzeichner die erste Bleistiftzeichnung
ANGEFERTIGT HAT. . . . After a master draughtsman *has
prepared* the first pencil drawing
—an example of the Perfect Tense which is dealt with in the next
chapter.

**Zeichnerin**: feminine form of **Zeichner**. Many such forms
exist in German. **Lehrerin** is another common example. The
plurals are **Zeichnerinnen, Lehrerinnen.**

## GRAMMAR

### Conjunctions

The words which link words or clauses together are called
conjunctions. For example, in

> *Jack oiled the thread AND the nut turned easily*
>     and
> *He tried to turn the screw BUT it was immovable*

the words *and* and *but* are conjunctions. In German there are
two sorts of conjunctions—

(*a*) Co-ordinating (which do not affect the normal word-
order, e.g. **Hans betrat das Zimmer ABER es war leer;
Er nahm das Buch UND schlug es auf**), and

(*b*) Subordinating (which introduce a subordinate clause, i.e.
they send the finite verb to the end of the clause which they
introduce, e.g. **Man glaubte, DAß die Erdölvorräte kaum
länger als 1½ Jahrzehnte reichen WÜRDEN. Diese Zeiten
sind lange vorbei, OBWOHL es eine Erdölindustrie erst
seit wenig mehr als** 100 **Jahren GIBT**).

The co-ordinating conjunctions are less numerous than the
others and so may be easily memorized. The rest, then, are
subordinating.

*Co-ordinating Conjunctions*

| | |
|---|---|
| **und** | *and* |
| **aber** | *but* |
| **allein** | *but* |

| sondern | *but* (after a negative, e.g. **Es war nicht Gold sondern Messing**) |
| oder | *or* |
| denn | *for* |

N.B. **Aber** does not always introduce or begin the clause. It may be used with the sense of *however*, e.g. **Der Arbeiter aber blieb stehen.**

*Subordinating Conjunctions*

| als | *when* | nachdem | *after* |
| bevor | *before* | ob | *whether, if* |
| ehe | *before* | obgleich | *although* |
| bis | *until* | obschon | *although* |
| da | *since* | obwohl | *although* |
| damit | *in order that* | seitdem | *since* |
| daß | *that* | während | *while* |
| falls | *in case* | wenn | *if, when, whenever* |
| indem | *while* | weil | *because* |

## WHEN

Three German words translate English *When:* **wenn, als** and **wann.**

**Wenn** with a present tense may mean *when* or *whenever—*

**Wenn es neun Uhr schlägt, sind alle Arbeiter in der Fabrik,** *When it strikes nine, all the workers are in the factory.*

**Wenn ich Zeit habe, lese ich Romane,** *Whenever I have time, I read novels.*

**Wenn** with a past tense can mean only *whenever—*

**Wenn er Ferien hatte, fuhr er immer nach Frankreich,** *Whenever he had a holiday he always went to France.*

**Als,** too, means *when*, but it refers only to one particular occasion and so can be used only with a past tense.

**Als der Arbeiter ankam, war die Fabrik geschlossen,**
*When the workman arrived, the factory was closed.*

**Wann**? is an interrogative word and so can only introduce a question, direct or indirect—

**Wann haben Sie Ferien?**  *When do you have your holidays?*

**Er fragte sie, wann sie Ferien hätte,** *He asked her when she had her holidays.*

# 11

# DER ZWECK DER ZEITSCHRIFT: STUDIUM GENERALE

*(Zeitschrift für die Einheit der Wissenschaften)*
Springer Verlag. Berlin, Göttingen, Heidelberg.

Die weite Entfaltung der Wissenschaft hat zwangsläufig zu einer Spezialisierung geführt, die einerseits ihre großen Erfolge ermöglicht, andererseits aber die Gefahr der Verengung des Horizontes und der Erstarrung in eingefahrenen Bahnen des Einzelfaches in sich trägt. Es ist notwendig, daß immer von neuem und auf verschiedensten Wegen eine Berührung und Begegnung der Fachwissenschaften herbeigeführt wird. Das STUDIUM GENERALE ist aus dieser Einsicht mit dem Ziel gegründet worden, den Blick auf die Einheit der Wissenschaft zurückzulenken, zwischen den Fachdisziplinen anregend zu vermitteln und damit einen Dienst für die Zusammenarbeit der Wissenschaften zu leisten. Die Diskussion über einzelne Begriffe, Problemkomplexe und Methoden, die viele Wissenschaftsgebiete angehen, soll dem Fachspezialisten die Möglichkeit geben, grundsätzliche Fragen über sein Fach hinaus weiter zu verfolgen und umgekehrt Anregungen für sein Fach zu erhalten.

Das STUDIUM GENERALE stellt auf diese Weise eine immer notwendiger werdende Ergänzung unserer fachwissenschaftlichen Literatur dar.

## VOCABULARY

| | | | |
|---|---|---|---|
| **der Zweck** | purpose, objective | **der Verlag** | publishing firm |
| **die Zeitschrift** | journal | **die Entfaltung** | unfolding, development |
| **die Einheit** | unity | | |
| **die Wissenschaft** | learning, scholarship, science | **die Spezialisierung** | specialization |

45

| der Erfolg | success |
|---|---|
| die Gefahr | danger |
| die Verengung | narrowing, limitation |
| der Horizont | horizon |
| die Erstarrung | stiffening, rigidi-fication |
| die Bahn | track |
| das Einzelfach | particular subject |
| der Weg | way |
| die Berührung | touching, making contact |
| die Begegnung | meeting, encoun-ter |
| die Fachwissen-schaft | special branch of study |
| die Einsicht | point of view |
| das Ziel | aim, intention |
| die Fachdiszip-lin | subject discipline |
| der Dienst | service |
| die Zusammen-arbeit | co-operation |
| der Begriff | idea, notion |
| der Problem-komplex | difficult problem, maze of prob-lems |
| die Methode | method |
| das Wissen-schaftsgebiet | sphere of know-ledge |
| der Fachspezia-list | subject specialist |
| die Möglichkeit | possibility |

| die Frage | question |
|---|---|
| das Fach | subject, line, speciality |
| die Anregung | suggestion, stimu-lus |
| die Weise | manner, way |
| die Ergänzung | supplement (com-pleting) |
| führen | lead |
| ermöglichen | render possible |
| tragen | carry, bear, wear |
| herbeiführen | bring about |
| gründen | found |
| zurücklenken | lead back |
| vermitteln | mediate |
| angehen | concern |
| geben | give |
| verfolgen | pursue, follow up |
| darstellen (Sep. verb) | represent |
| weit | considerable, extensive |
| zwangsläufig | inevitably |
| einerseits | on the one hand |
| andererseits | on the other hand |
| eingefahren | well-worn, con-ventional |
| von neuem | anew |
| anregend | interestingly |
| einzeln | particular |
| grundsätzlich | fundamental |
| umgekehrt | vice versa |

*For special note*

| hat . . . geführt | has led |
|---|---|
| ist . . . gegründet worden | has been founded |

(These are examples of the Perfect Tense, for which see the following Grammar Section)

**Erstarrung in eingefahrenen Bahnen des Einzelfaches** "getting stuck in the rut of one's speciality."

| den Blick lenken | direct the attention |
|---|---|
| einen Dienst leisten | perform a service |

**immer notwendiger werdende Ergänzung** a completion which is becoming daily more necessary. **werdend,** becoming; present participle. (*See* Grammar Section below.)

| | |
|---|---|
| **stellt . . . dar** | from **darstellen,** represent |
| **auf diese Weise** | in this way |

### GRAMMAR

In English the *Perfect Tense* is formed by adding the Past Participle of the verb to a finite form of the verb *to have*, e.g.

*take:*     (*he*) *has taken*
        (*they*) *have taken*

In former times the perfect tense of some verbs was formed using the verb *to be* as the auxiliary, e.g.

*come:*     (*he*) *is come*
        (*they*) *are come*

In German both ways are still current.

Most German verbs form their Perfect with finite forms of **haben,** e.g.

| | |
|---|---|
| **Ich habe gemacht** | *I have made* |
| **er )** | *he )* |
| **sie) hat gemacht** | *she) has made* |
| **es )** | *it )* |
| **wir haben gemacht** | *we have made* |
| **Sie haben gemacht** | *you have made* |
| **sie haben gemacht** | *they have made* |

Others form their Perfect with **sein** and these should be specially learnt, e.g. **kommen, gehen, sein, fallen, werden.**

| | |
|---|---|
| **ich bin gekommen** | *I have come* |
| **er )** | *he )* |
| **sie) ist gekommen** | *she) has come* |
| **es )** | *it )* |
| **wir sind gekommen** | *we have come* |
| **Sie sind gekommen** | *you have come* |
| **sie sind gekommen** | *they have come* |

The past participle of German weak verbs is formed by adding **ge-** as a prefix to the stem of the verb and **-t** as a suffix, e. g.

**machen**
**ge–mach–t**

The past participle of German strong verbs has to be learnt with the "principal parts".* Besides having the prefix **ge-** it usually has **–en** as its suffix.

Here are some examples—

| | |
|---|---|
| **nehmen** (*take*) | p.p. **genommen** |
| **gehen** (*go*) | p.p. **gegangen** |
| **treiben** (*drive*) | p.p. **getrieben** |

Verbs with separable prefixes (see p. 23) retain the prefix in the past participle, but it is followed by the **ge** and the usual stem and ending, e.g.

| | |
|---|---|
| **zumachen** (*close*) | p.p. **zugemacht** |
| **zurückkommen** (*return*) | p.p. **zurückgekommen** |

Verbs with inseparable prefixes do not have **ge** in their past participles, e.g.

**verbessern** (*improve*)    p.p. **verbessert**

*The Present Participle*

In the following English sentences the words ending in *-ing* are present participles—

*He bought a* singing *bird*
Shouting *at the top of his voice he leapt into the water*
*We shall be* waiting *for you*

In English we use the present participles of verbs a great deal. In German they are used much less frequently and for one purpose only, i.e. as adjectives and adverbs, e.g.

**ein TREFFENDES Beispiel,** *a* striking *example*
**ANREGEND zu vermitteln,** *to* convey *interestingly*

In German the present participle is made by adding **d** to the infinitive of the verb, e.g. **singen-D, rotieren-D.**

* See list, p. 154 *et seq.*

# 12

# ÖLHYDRAULIK

Hydraulische Antriebe werden heute in besonders rasch zunehmendem Maße auf allen Gebieten des Maschinenbaues verwendet. Die „Ölhydraulik" wendet sich deshalb in erster Linie an alle Ingenieure des allgemeinen Maschinenbaues, die mit der Entwicklung von Maschinen zu tun haben, in denen hydraulische Antriebe eingeführt werden sollen oder bereits verwendet werden.

Zahlreiche Tabellen, Schaltpläne für die Lösung der verschiedensten Bewegungsaufgaben, Schnittzeichnungen sowie die Zusammenstellung der wichtigsten physikalischen Grundlagen auf den Gebieten der Hydrodynamik, Hydrostatik und Thermodynamik, soweit diese Gebiete für die Hydraulik-Ingenieure von Bedeutung sind, machen sie aber auch zu einem wertvollen Nachschlagewerk für Fachingenieure auf dem Gebiet der Ölhydraulik.

Alle bisher verfügbaren Normbauteile der Ölhydraulik, wie Pumpen, Steuerschieber, Druckventile, Regelventile, Stromregler usw., werden ausführlich besprochen. Eine ganze Reihe von Anwendungsgebieten der Hydraulik wird an Hand von Beispielen behandelt.

Der Aufbau des Buches ermöglicht auch dem theoretisch weniger Vorgebildeten eine rasche Einarbeitung in jene Arbeitsgebiete der Hydraulik, die bisher als besonders schwierig betrachtet wurden, wie etwa das Gebiet der nichtstationären Strömungsvorgänge und Druckschwingungen.

ZOEBL H. „Ölhydraulik" (Wien 1963, Springer Verlag)

## VOCABULARY

| | | | |
|---|---|---|---|
| der Antrieb | drive | der Aufbau | construction |
| das Maß | measure | der Vorgebil- | trained person |
| das Gebiet | province, sphere | dete | |
| der Maschinen- | machine con- | die Einarbeitung | introduction (to) |
| bau | struction | (in) | |
| die Ölhydraulik | oil hydraulics | der Strömungs- | flow process |
| die Tabelle | table, list | vorgang | |
| der Schaltplan | circuit diagram | die Druck- | variation of pres- |
| die Lösung | solution, solving | schwingung | sure |
| die Bewegungs- | problem of move- | | |
| aufgabe | ment | verwenden | apply, use |
| die Schnitt- | sectional diagram | wenden | turn |
| zeichnung | | einführen | introduce |
| die Zusammen- | summary, list | besprechen | discuss |
| stellung | | behandeln | treat |
| die Grundlage | principle | betrachten | consider |
| die Hydro- | hydrodynamics | | |
| dynamik | | hydraulisch | hydraulic |
| die Hydrostatik | hydrostatics | zunehmend | increasing |
| die Thermo- | thermodynamics | alle | all |
| dynamik | | deshalb | therefore |
| die Bedeutung | significance | bereits | already |
| das Nach- | work of reference | zahlreich | numerous |
| schlagewerk | | sowie | as well as |
| der Normbau- | standard compo- | wichtig | important |
| teil | nent | physikalisch | physical |
| die Pumpe | pump | wertvoll | valuable |
| der Steuer- | control valve | soweit | in so far as |
| schieber | | bisher | up to now. previ- |
| das Druckventil | pressure or safety | | ously |
| | valve | verfügbar | available |
| das Regelventil | bleeder-valve | ausführlich | in detail |
| der Stromregler | flow-regulator | schwierig | difficult |
| die Reihe | series | etwa | perhaps |
| die Anwendung | application | nichtstationär | non-stationary, |
| | | | unsteady |

*For special note*

**an Hand von Beispielen**                    with the help of examples

## GRAMMAR

### REFLEXIVE VERBS

Some verbs have an object which is the same person or thing as their subject. For example, in *I* wash *myself*, *I* and *myself* are the same person. The verb in this sentence is said to be used reflexively, i.e. the object is a reflexion, as it were, of the subject. The corresponding German forms are—

| | |
|---|---|
| **Ich wasche MICH** | . . . *myself* |
| **Er** ) | |
| **Sie) wäscht SICH** | . . . *himself, herself, itself* |
| **Es** ) | |
| **Wir waschen UNS** | . . . *ourselves* |
| **Sie waschen SICH** | . . . *yourself* |
| **sie waschen SICH** | . . . *themselves* |

N.B. Some verbs are reflexive in German which are not so in English. In Lesson 4 we had, for instance, **sich drehen,** *to turn, revolve*—

**Der Zylinder dreht sich,** *the cylinder turns (revolves)*

and in this lesson we have **sich wenden,** also *to turn*—

**Die „Ölhydraulik" wendet sich an alle Ingenieure,**
*"Oil Hydraulics" turns to* (or, *is aimed at*) *all engineers.*

The reflexive verb often translates an English passive construction—

**Eine Drehbank BEFINDET SICH in jeder Werkstätte,**
*A lathe is found in every workshop.*

# 13

## DER KLEINWAGEN „SPATZ"

*Was Sie an technischen Einzelheiten interessieren wird*

Der Motor ist ein 191-ccm-Einzylinder-F & S-Zweitakter mit Gebläsekühlung, der 10,2 PS schafft. Auch wenn Sie keine Garage haben, steht er im Winter im Freien, und haben Sie ihn gut zugedeckt, dann wird er Sie nicht enttäuschen. Die Bohrung beträgt 65 mm, der Hub 58 mm. Der Tank enthält 12 Liter, mit denen man normalerweise über 250 km fahren kann, denn der Normverbrauch (100 km auf ebener Straße bei gleichbleibenden 55 km/h) ist 4 Liter. In der Spitze fährt der Spatz 75 km/h. Eine Vierscheiben-Lamellenkupplung überträgt die Kraft auf ein Vierganggetriebe im Motorblock. Der Motor ist mit einer Dyna-Startanlage von 12 Volt mit 90–135 Watt ausgerüstet, die für Vor- und Rückwärtslauf eingerichtet ist. Die Zahnstangenlenkung ermöglicht als kleinsten Wendekreis ca. 9,50 m. Die auf vier Räder wirkende Öldruckbremse mit 180-mm-$\phi$-Bremstrommeln bringt den Wagen auf kürzeste Entfernung zum Stehen. Die als Seilzugbremse ausgeführte Handbremse wirkt auf die Vorderräder. Die Bereifung ist 4,40 × 12 auf Stahlscheiben-Rädern mit aufgesetzten Zierkappen. Der Spatz ist 3400 mm lang und 1450 mm breit, sein Radstand ist 1950 mm, die Spurweite 1160 mm, die Bodenfreiheit 185 mm. Sein Leergewicht beträgt 320 kg, und er kann 240 kg mitnehmen. Rahmen: stabile Zentralrohr-Konstruktion. Die Vorderachse ist einzeln in Lenkern aufgehängt und mit hydraulisch gedämpften Federbeinen abgefedert. Hinten befindet sich eine quergelenkte Pendelachse mit einzeln aufgehängten Rädern, die auch mit Federbeinen und hydraulischen Stoßdämpfern abgefedert ist. Das Finanzamt verlangt im Jahr vom Spatz-Besitzer nur 29.-DM, die Haftpflichtversicherung beträgt 90.-DM.

Änderungen im Laufe der ständigen, fortschreitenden Entwicklung vorbehalten.

## VOCABULARY

| | | | |
|---|---|---|---|
| **der Kleinwagen** | miniature car, "mini"-car | **PS= Pferde-stärke** | horse-power |
| **der Spatz** | sparrow | **das Freie** | the open air |
| **der Motor** | motor, engine | **die Bohrung** | bore |
| **der Zweitakter** | two-stroke (engine) | **der Hub** | stroke (of the piston) |
| **die Gebläseküh-lung** | fan-cooling | **der Normver-brauch** | normal consumption |
| | | **die Spitze** | maximum |

**die Vierscheiben-Lamellenkupp-lung**   four disc laminated clutch
**das Vierganggetriebe**   four-speed gear
**der Motorblock**   engine block

| die Startanlage | starter mechanism |
| der Vor- und Rückwärtslauf | forward and backward movement |
| die Zahnstangenlenkung | rack steering |

| die Wendekreis | turning circle | die Bereifung | tyres, tyring |
| die Öldruck-bremse | oil pressure brake | die Zierkappe | hub-cover |
| | | der Radstand | wheel base |
| die Bremstrom-mel | brake drum | die Spurweite | track width |
| | | die Bodenfrei-heit | ground clearance |
| die Entfernung | distance | | |
| die Seilzug-bremse | cable brake | das Leergewicht | weight when empty |
| die Handbremse | hand brake | der Rahmen | frame |
| das Vorderrad | front wheel | | |

| die Zentralrohr-Konstruktion | central tube construction |

| die Vorderachse | front axle | bringen | bring |
| der Lenker | link, linkage | wirken | operate, work |
| das Federbein | spring leg, tele-scopic leg | mitnehmen | take (with), carry |
| | | abfedern | spring |
| die Pendelachse | swing axle | befinden (sich) | be (situated) |
| das Finanzamt | Treasury | verlangen | demand |
| der Stoßdämp-fer | shock absorber | | |
| | | normalerweise | normally |
| der Besitzer | owner | denn (*conjunction*) | for |
| die Haftpflicht-versicherung | third party insur-ance | eben | level |
| | | gleichbleibend | steady |
| DM. (Deutsche Mark) | German Mark | ca. (*circa*) | about |
| | | aufgesetzt | applied, pushed on |
| die Änderung | alteration, change | stabil | sturdy |
| der Lauf | course, movement | einzeln | independently |
| | | aufgehängt | suspended |
| schaffen | create, develop | hydraulisch ge-dämpft | hydraulically "smoothed" or "damped" |
| stehen | stand | | |
| zudecken | cover | | |
| enttäuschen | disappoint | hinten | behind |
| betragen | amount to | quergelenkt | cross-braced |
| enthalten | contain, hold | ständig | constant |
| übertragen | transmit | fortschreitend | progressive |
| ausrüsten | equip | vorbehalten | reserved |
| einrichten | furnish, equip | | |

*For special note*

die als Seilzugbremse ausgeführte Handbremse, *the handbrake operated by a cable*

**auf kürzeste Entfernung zum Stehen,** *to a standstill in the shortest distance*

A **Dyna-Startanlage** serves as both dynamo and starter.

## NUMBERS—FRACTIONS—DECIMALS

| | | | |
|---|---|---|---|
| 0 | **null** | 19 | **neunzehn** |
| 1 | **eins** | 20 | **zwanzig** |
| 2 | **zwei** | 21 | **ein und zwanzig** |
| 3 | **drei** | 22 | **zwei und zwanzig,** |
| 4 | **vier** | | etc. |
| 5 | **fünf** | 30 | **dreißig** |
| 6 | **sechs** | 40 | **vierzig** |
| 7 | **sieben** | 50 | **fünfzig** |
| 8 | **acht** | 60 | **sechzig** |
| 9 | **neun** | 70 | **siebzig** |
| 10 | **zehn** | 80 | **achtzig** |
| 11 | **elf** | 90 | **neunzig** |
| 12 | **zwölf** | 100 | **hundert** |
| 13 | **dreizehn** | 101 | **hundert (und) eins** |
| 14 | **vierzehn** | 200 | **zwei hundert** |
| 15 | **fünfzehn** | 1,000 | **tausend** |
| 16 | **sechzehn** | 1,000,000 | **eine Million** |
| 17 | **siebzehn** | 1,000,000,000 | **eine Milliarde** |
| 18 | **achtzehn** | | |

*Ordinal Numbers*

| | | | |
|---|---|---|---|
| | **der nullte** | 19th | **der neunzehnte** |
| 1st | **der erste** | 20th | **der zwanzigste** |
| 2nd | **der zweite** | 30th | **der dreißigste** |
| 3rd | **der dritte** | 100th | **der hundertste** |
| 4th | **der vierte** | 1000th | **der tausendste** |
| 7th | **der siebte** | 1,000,000th | **der millionste** |

N.B. These words are adjectives and so "agree" in gender, case and number with the nouns which follow—

**der erste Mann, die zweite Frau, das vierte Kind, im ersten Augenblick**

| **einfach** | | *onefold*, or *simple* | |
| **zweifach** | | *twofold* | |
| **dreifach,** etc. | | *threefold* | |

| **einmal** | *once* | **hundertmal** | 100 *times* |
| **zweimal** | *twice* | **tausendmal** | 1,000 *times* |
| **dreimal** | *thrice* | | |
| **viermal,** etc. | *four times* | | |

| **erstens** | *in the first place* |
| **zweitens** | *secondly* |
| **drittens,** etc. | *thirdly* |

| **zuerst** | *at first* |
| **zu–zweit, zu zweien** | *secondly* |
| **zu–dritt, zu dreien** | *thirdly* |
| **zu–viert, zu vieren** | *fourthly* |
| **zu–zehnt** | *tenthly* |
| **zu–zwanzig** | *in the twentieth place* |

## Fractions

| $\frac{1}{2}$ | **die Hälfte** | or **ein halb (er)** |
| $\frac{1}{4}$ | **ein Viertel** | |
| $\frac{2}{3}$ | **zwei Drittel** | |
| 1/100 | **ein Hundertstel** | |
| 3/1,000 | **drei Tausendstel** | |
| 1/1,000,000 | **ein Millionstel** | |

N.B.

| $1\frac{1}{2}$ | **anderthalb** |
| $2\frac{1}{2}$ | **zwei (und) einhalb** |
| $3\frac{1}{2}$ | **drei (und) einhalb** |

## Decimals

The decimal point is expressed by a comma—

| 0,1 | **nullkommaeins** |
| 0,01 | **nullkommanulleins** |
| $\pi$ (Pi) 3,14159 | **dreikommaeinsviereinsfünfneun** |

*Mathematical Symbols*

| | | |
|---|---|---|
| $+$ | **plus (und)** | |
| $-$ | **minus (weniger)** | |
| . (or $\times$) | **mal** | *times* |
| : (or /) | **geteilt durch** | *divided by* |
| $=$ | **gleich** | |
| $\equiv$ | **identisch** | |
| $>$ | **größer als** | *greater than* |
| $<$ | **kleiner als** | *less than* |
| $\sqrt[n]{\ }$ | **n<sup>te</sup> Wurzel aus** | |
| log. | **Logarithmus** | |
| $\infty$ | **unendlich** | *infinity* |
| $x^2$ | **die zweite Potenz** | |
| $x^3$ | **die dritte Potenz** | |
| $x^4$ | **die vierte Potenz** | |

*to cube,* **kubieren, zur dritten Potenz erheben**
*the cube root:* **die Kubikwurzel.**

# 14

## FELIX WANKELS
## REVOLUTIONÄRER MOTOR

Viele Jahre lang hegte ein deutscher Erfinder namens Felix Wankel einen revolutionären Gedanken: Er wollte den klassischen Verbrennungsmotor durch etwas besseres ersetzen. Vor drei Jahren war es so weit. Die NSU-Werke bauten nach seiner Idee den Wankelmotor. Dieser Motor erregte großes Aufsehen.

Der übliche Verbrennungsmotor mit sich auf und abbewegenden Hubkolben steht heute am Ende seiner Entwicklungsmöglichkeiten. Man mag den ölglatten, fast geräuschlosen Kraftfluß in einem feinen Wagen bewundern. Man bewundert die Geschicklichkeit der Techniker, die ihn auf diesen Entwicklungsstand brachten, aber er hat seine Nachteile, und etwas besseres ist möglich.

„Sich auf und ab bewegend," hier liegen die Grenzen des konventionellen Hubkolbenmotors. Eine solche Maschine ist im Effekt wie ein vielrohriges Geschütz, in dem das Pulver durch ein Benzinluftgemisch, die Granaten durch die Kolben und die Zünder durch die Zündkerzen ersetzt sind. Das Gemisch wird gezündet, dehnt sich explosionsartig aus, die mit den Pleuelstangen verbundenen Kolben suchen aus den „Rohren," den Zylindern, herauszufliegen, müssen aber schon nach wenigen Zentimetern Kolbenweg angehalten und in umgekehrter Richtung zurückgeführt werden. Dieser ständige, in jeder Minute des Motorlaufs sich einige tausendmal wiederholende Wechsel in der Bewegungsrichtung der Kolben ist technisch im Grunde zweitrangig.

(Auszug aus der Monatschrift *Das Beste aus Reader's Digest*, Oktober, 1960)

## VOCABULARY

| | |
|---|---|
| **der Gedanke** | thought, idea |
| **der Verbren-** | internal combus- |
| **nungsmotor** | tion engine |
| **die Idee** | idea |
| **das Aufsehen** | stir, sensation |
| **der Kolben** | piston |
| **die Entwick-** | possibility of |
| **lungsmöglich-** | development |
| **keit** | |
| **der Kraftfluß** | flow of power |
| **die Geschick-** | cleverness, skill |
| **lichkeit** | |
| **der Entwick-** | stage of develop- |
| **lungsstand** | ment |
| **der Nachteil** | disadvantage |
| **die Grenze** | border, limit |
| **der Hubkolben-** | piston-engine |
| **motor** | |
| **der Effekt** | effect |
| **das Geschütz** | gun |
| **das Pulver** | powder |
| **das Benzinluft-** | petrol–air-mixture |
| **gemisch** | |
| **die Granate** | shell |
| **der Zünder** | detonator |
| **die Zündkerze** | spark-plug |
| **das Gemisch** | mixture |
| **die Pleuelstange** | connecting rod |
| **das Rohr** | tube, pipe, barrel |
| | (of gun) |
| **der Kolbenweg** | path of the piston |
| **die Richtung** | direction |
| **die Minute** | minute |
| **der Motorlauf** | running of the |
| | motor |
| **der Wechsel** | change, alteration |
| **der Grund** | basis |

| | |
|---|---|
| **der Auszug** | extract |
| **die Monat-** | monthly periodi- |
| **schrift** | cal |
| **hegen** | cherish |
| **bauen** | build, construct |
| **erregen** | excite |
| **bewundern** | admire |
| **bringen** (p.t. | bring |
| **brachte**) | |
| **zünden** | ignite |
| **ausdehnen** (sep.) | expand |
| **(sich)** | |
| **suchen** | seek, try |
| **herausfliegen** | fly out |
| **(sep.)** | |
| **anhalten** (sep.) | stop, check |
| **zurückführen** | lead back |
| **wiederholen** | repeat |
| **namens** | by name |
| **revolutionär** | revolutionary |
| **klassisch** | classical |
| **üblich** | usual |
| **ölglatt** | oil-smooth |
| **fast** | almost |
| **geräuschlos** | noiseless |
| **fein** | elegant, fine |
| **möglich** | possible |
| **konventionell** | conventional |
| **solch** | such |
| **vielrohrig** | of many barrels |
| **explosionsartig** | like an explosion, |
| | explosively |
| **verbunden** | connected |
| (from **verbinden**) | |
| **zweitrangig** | second-rate |

## Special points

**etwas Besseres,** something better

The adjective which follows **etwas** always agrees with it in being neuter: **etwas Gutes, etwas Schönes.**

**war es so weit,** it had reached this stage

**nach seiner Idee,** in accordance with his idea

**mit sich auf und abbewegenden Hubkolben,** with pistons moving up and down, or backwards and forwards

**Man mag . . . bewundern,** One may . . . admire

**eine solche Maschine,** *such a* machine—note the order in German

**die mit den Pleuelstangen verbundenen Kolben,** the pistons connected with the connecting rods

**Dieser . . . sich einige tausendmal wiederholende Wechsel,** This . . . change repeating itself some thousands of times

**im Grunde,** basically

## GRAMMAR

### Inseparable Prefixes

Some verbs have prefixes which are not separable. In this lesson we have examples in **erregen, bewundern, ersetzen, verbinden.**

The following prefixes are always inseparable: **be-, emp-, er-, ge-, hinter-, miß-, ver-, voll-, wider-, zer-.**

When a verb has such an inseparable prefix it does not have **ge-** prefixed to it in the past participle.

**Wankel hat grosses Aufsehen ERREGT**

**Der Kolben ist mit der Pleuelstange VERBUNDEN**

# 15

# DAS MOPEDFAHREN. KURZE ANWEISUNG FÜR DIE ERSTE FAHRT

*A. Vorbereitungen*

Jeder SACHS-Motor ist schon im Werk einige Zeit gelaufen. Nicht nur auf dem Prüfstand, auch auf der Straße wurde er probiert. Der Motor ist also fahrbereit und das Getriebe mit Öl gefüllt. Sie brauchen nur noch Kraftstoff zu tanken und den Reifendruck zu prüfen.

Vergewissern Sie sich auch, daß die Luftlöcher in der Öleinfüllschraube auf dem Getriebe frei und nicht noch vom Transport her durch einen Klebestreifen verschlossen sind. Sonst kann im Betrieb leicht Getriebeöl am Schalthebel herausgedrückt werden.

*B. Tanken:* Zweitakt-Gemisch 1:25

25 Teile normalen Markenbenzins werden mit 1 Teil Motorenöl Zähigkeit SAE 50, am besten „SACHS-Motor-Öl," in einem besonderen Gefäß gut vermischt. Auf 2 Ltr. Benzin nimmt man also 80 ccm Öl. Keine Öle mit Zusätzen (legierte oder HD-Öle) und keine sogenannten gefetteten Öle (Rennöle) verwenden, nur reines Mineralöl verlangen! Sie fahren damit billiger und mindestens ebenso gut.

*C. Bedienungshebel am Sachs-Moped*

1. *Gasdrehgriff.* Durch Drehen nach hinten wird Gasschieber im Vergaser geöffnet.

2. *Schaltdrehgriff mit Kupplungshebel* links am Lenker. Wird der Hebel gezogen, trennt die Kupplung den Kraftfluß vom Motor zum Getriebe und Hinterrad. Gleichzeitig wird die Verriegelung des Schaltdrehgriffs aufgehoben, der Griff mit

Kupplungshebel kann jetzt nach oben und unten geschwenkt und dadurch der gewünschte Gang oder der Leerlauf eingeschaltet werden.

3. *Handbremshebel* rechts am Lenker. Wirkt auf Vorderradbremse.

4. *Kurzschlußknopf* im Scheinwerfer zum Ausschalten der Zündung.

5. *Kraftstoffhahn* am Tank.

6. *Tupfer am Vergaser.*

## D. Anfahren

1. Kraftstoffhahn öffnen.

2. Bei kaltem Motor Tupfer am Vergaser 5 bis 6 Sekunden ruhig herunterdrücken.

3. Kein Gas geben.

4. Ersten Gang einschalten. Hierzu Kupplungshebel ziehen und Schaltgriff so nach vorn unten drehen, daß Marke am Griffstück auf 1 zeigt.

5. Mit gezogener Kupplung anfahren.

6. Kupplung langsam loslassen, dabei weitertreten bis Motor anspringt.

7. Dann erst Gas geben.

8. Wenn Motor nach etwa 10 m nicht angesprungen ist, etwas mit Gasdrehgriff spielen. Bleibt der Motor wieder stehen, nochmals tupfen.

## VOCABULARY

| | | | |
|---|---|---|---|
| **die Vorberei-tung** | preparation | **die Öleinfüll-schraube** | oil filling cap |
| **die Anweisung** | instruction(s) | **das Getriebe** | gear box |
| **die Fahrt** | journey, trip | **der Klebestreif-en** | adhesive strip |
| **das Werk** | work, factory, works | **der Betrieb** | operation, working |
| **der Prüfstand** | testing stand | | |
| **das Öl** | oil | **das Getriebeöl** | gear-oil |
| **der Kraftstoff** | fuel | **der Schalthebel** | gear lever |
| **der Reifendruck** | tyre pressure | **das Zweitakt-Gemisch** | two-stroke mixture |
| **das Luftloch (-löcher)** | air-hole | **das Benzin** | petrol |

| | | | |
|---|---|---|---|
| das Marken-benzin | proprietary petrol | tanken | put into tank |
| die Zähigkeit | viscosity | prüfen | test |
| das Gefäß | vessel | vergewissern | assure, make sure |
| der Zusatz | additive | verschließen | shut |
| das Rennöl | racing oil | (p.p. verschlos-sen) | |
| der Bedienungs-hebel | operating lever | herausdrücken | push out |
| der Gasdrehgriff | twist grip | vermischen | mix |
| der Gasschieber | fuel control valve | nehmen | take |
| der Vergaser | carburettor | öffnen | open |
| der Schaltdreh-griff | twist grip for gear changing | ziehen | pull |
| der Kupplungs-hebel | clutch lever | trennen | separate |
| der Hebel | lever | aufheben | release |
| die Kupplung | clutch | schwenken | turn |
| das Hinterrad | rear wheel | einschalten | engage |
| die Verriegelung | locking | anfahren | start off |
| der Griff | grip | herunterdrücken | press down |
| der Gang | gear | loslassen | let go |
| der Leerlauf | neutral gear | weitertreten | paddle forwards |
| der Handbrems-hebel | handbrake lever | anspringen | start up |
| die Vorderrad-bremse | front wheel brake | spielen | play, experiment |
| der Kurzschluß-knopf | short circuiting switch-button | tupfen | use the **Tupfer** (see above), "tickle" |
| der Schein-werfer | headlamp | fahrbereit | ready for the road |
| das Ausschalten | switching-off | frei | free |
| die Zündung | ignition | sonst | otherwise |
| der Kraftstoff-hahn | fuel cock | besonder | special |
| der Tupfer | carburettor opera-ting rod, "tickler" | legiert | mixed, compoun-ded |
| das Anfahren | starting off | sogenannt | so-called |
| der Schaltgriff | gear-change grip | gefettet | enriched |
| die Marke | mark | rein | pure |
| das Griffstück | grip | ebenso | just as |
| | | links | on the left |
| probieren | test | gleichzeitig | simultaneous(ly) |
| | | nach oben | upwards |
| | | nach unten | downwards |
| | | gewünscht | desired |
| | | rechts | on the right |
| | | kalt | cold |
| | | ruhig | gently |

| **hierzu** | for that purpose | **gezogen** (p.p. of | drawn, lifted, dis- |
| **nach vorn(e)** | forwards | **ziehen)** | engaged |
| **unten** | down | **nochmals** | again, once more |

## GRAMMAR

### IMPERATIVES

**Vergewissern Sie sich, daß die Luftlöcher frei sind,** *Make sure* (lit: *Assure yourself*) *that the air holes are free.*

**Keine Öle mit Zusätzen VERWENDEN, nur reines Mineralöl VERLANGEN!** *Employ no oils with additives; demand pure mineral oil only!*

Two ways of giving orders or instructions are employed here.

(*a*) **Vergewissern Sie sich,** employing the pronoun **Sie,** is the imperative proper. Here are some more examples—

| **Kommen Sie hier!** | *Come here!* |
| **Bleiben Sie stehen!** | *Stop!* |
| **Setzen Sie sich!** | *Sit down!* |

(*b*) The infinitive is used with imperative meaning. It is placed at the end of the sentence—

| **Kein Gas geben.** | *Give no gas.* |
| **Kupplung langsam loslassen.** | *Release clutch slowly.* |

# 16

# KUGELLAGER

Aus frühesten Zeiten sind Hilfsmittel bekannt, die als Vorläufer des Kugel- und Rollenlagers gelten können: Im alten Ninive wurden schwere Steinblöcke über Holzwalzen geschoben. Sklaven trugen die hinten frei gewordenen Rollen nach vorne, um sie neu auszulegen. Um 330 v.Chr. erfand Diades ein

die Kugel — — die Kugel

der Innenring — — der Außenring

Kugellager

Rollenlager für Belagerungsmaschinen, und aus der Zeit des Kaisers Caligula kennt man ein Kugellager, auf dem ein drehbares Standbild ruhte.

Wohl als erster befaßte sich kein Geringerer als Leonardo da Vinci mit der wissenschaftlichen Untersuchung der Reibung. Ein erstes Patent auf Radialkugellagern für Wagenachsen konnte dem Engländer Vaughan 1794 erteilt werden. Noch lange gaben Tragfähigkeit, Bauart, Werkstoff usw. hemmende Probleme auf. Durch die allgemeine Industrialisierung, besonders durch das Erscheinen des Automobils um die Jahrhundertwende, stellte sich der Wunsch nach einem allen Anforderungen gerecht werdenden Wälzlager immer dringender. Der deutsche Professor Stribeck schuf zunächst durch grundlegende und noch heute gültige wissenschaftliche Untersuchungen die Basis zu dessen Verwirklichung.

1907 gelang dem Schweden Wingquist, Gründer der weltbekannten SKF-Werke, die Erfindung des Pendelkugellagers, das so gestaltet ist, daß es sich bei Schräglage der Welle selbsttätig bis zu einem gewissen Grad einstellen kann. Ein Jahr später verwirklichte Dr. Kirner in Deutschland ein brauchbares Zylinderrollenlager, das sich besonders dort gut eignet, wo schwere Belastungen erhöhte Tragfähigkeit voraussetzen. Heute bietet SKF mehrere tausend Lagerarten an: von 10 bis 1400 mm Außendurchmesser, von 1,5 g bis fast 2,5 t Gewicht.

Aus „Große Erfindungen" (Bern, Schweiz, Verlag
Hallweg)

## VOCABULARY

| | | | |
|---|---|---|---|
| **das Hilfsmittel** | remedy, expedient, aid | **die Jahrhundertwende** | turn of the century |
| **der Vorläufer** | precursor | **der Wunsch** | wish |
| **das Kugellager** | ball bearing | **die Anforderung** | demand |
| **das Rollenlager** | roller bearing | **das Wälzlager** | roller bearing |
| **der Steinblock (blöcke)** | block of stone | **die Verwirklichung** | realization |
| **die Holzwalze** | wooden roller | **der Gründer** | founder |
| **der Sklave** | slave | **das Pendelkugellager** | self-aligning ball bearing |
| **die Belagerung** | siege | | |
| **der Kaiser** | emperor | **die Schräglage** | slanting position |
| **das Standbild** | statue | **die Welle** | shaft |
| **die Untersuchung** | investigation | **der Grad** | degree |
| | | **das Zylinderrollenlager** | parallel roller bearing |
| **die Reibung** | friction | | |
| **das Patent** | patent | **die Belastung** | loading |
| **die Tragfähigkeit** | carrying capacity | **die Lagerart** | kind of bearing |
| **die Bauart** | construction | **der Außendurchmesser** | outside diameter |
| **die Industrialisierung** | industrialization | **schieben** (p.p. **geschoben**) | push, shove |
| **das Erscheinen** | appearance | **auslegen** | lay out |

| | | | |
|---|---|---|---|
| **ruhen** | rest | **bekannt** | known |
| **sich befassen** | occupy, busy one-self | **schwer** | heavy |
| | | **v.Chr. (= vor Christus)** | B.C. |
| **erteilen** (p.p. erteilt) | accord, grant | **drehbar** | rotatory |
| **aufgeben** | give up, set (problem) | **wohl** | well, probably |
| | | **gering** | little |
| **hemmen** | hinder, hamper | **kein Geringerer** | no less a person |
| **sich stellen** | pose itself, become | **wissenschaftlich** | scientific |
| **dringen** | press, urge | **gerecht werdend** | satisfying |
| **gelingen** (p.t. gelang) | succeed (*see* grammar) | **zunächst** | first |
| **gestalten** | form | **grundlegend** | fundamental |
| **einstellen** | regulate, align, adjust | **gültig** | valid |
| | | **weltbekannt** | known throughout the world |
| **verwirklichen** | realize, achieve | **selbsttätig** | automatically |
| **sich eignen** | fit, be suitable | **brauchbar** | practicable |
| **voraussetzen** | assume, presuppose | **erhöht** | increased |
| | | **mehrere** | several |
| **anbieten** | offer | | |

*Points for special note*

**die als Vorläufer . . . gelten können,** which may be regarded as forerunners

**stellte sich der Wunsch immer dringender,** the wish was expressed more and more urgently

**nach einem allen Anforderungen gerecht werdenden Wälzlager,** for a roller-bearing satisfying all requirements

**1907 gelang dem Schweden Wingquist die Erfindung,** in 1907 the Swede, Wingquist, succeeded in inventing (*see* Grammar below).

### GRAMMAR

#### IMPERSONAL USE OF VERBS

When we say *It is raining*, the pronoun *it* refers to no person or thing in particular. The verb rain in *it rains* is said to be used impersonally. In German "impersonal verbs" are more common.

The verb **gelingen** used in the passage above is common because it is the normal way of expressing one's success in doing something. **Es gelingt mir, ihm, ihr, ihnen,** means literally *It succeeds to me, to him, to her, to them* but will be translated: *I succeed, he succeeds,* etc.

*I succeed in breaking the bar* is translated: **Es gelingt mir, die Stange ZU BRECHEN.**

Other commonly occurring impersonal uses are these—

| | |
|---|---|
| **Es friert mich,** or **Mich friert** | *I am cold* |
| **Es freut mich** | *I rejoice, I am glad* |
| **Es glückt mir** | *I succeed* |
| **Es mißlingt mir** | *I am unsuccessful, I fail* |
| **Es läßt sich wünschen** | *It is desirable* |

# 17

# DRUCKLUFTKOMPRESSOREN FÜR DIE STEINGEWINNUNG UND —BEARBEITUNG

Von den heute gebräuchlichen Energien, die zur Arbeitsleistung ausgenützt werden, wird Druckluftenergie hauptsächlich für Arbeitsverfahren angewandt, bei denen erhöhte Forderungen der allgemeinen Sicherheit und der Sicherheit für das Bedienungspersonal sowie sehr hohe Forderungen an die Leistungsfähigkeit der Arbeitsmaschinen gestellt werden. Es ist deshalb nicht verwunderlich, wenn bei Handwerkzeugen Druckluftenergie vorherrscht, weil die hohen Lohn- und Gemeinkosten wirtschaftliche Betrachtungen maßgebend beeinflussen und man aus diesem Grunde bestrebt ist, die Arbeitsleistung dadurch zu steigern, daß man möglichst leistungsfähige Werkzeuge zur Verfügung stellt.

Die Drucklufterzeugung geschieht durch Kompressoren, und es gibt in Deutschland eine sehr leistungsfähige Industrie, die sich mit der Konstruktion und Fertigung von Druckluftkompressoren beschäftigt. Man trifft je nach den Anwendungsgebieten sehr verschiedene Bauarten von Kompressoren an. Während sich die Strömungsmaschinen nur für sehr große Luftmengen eignen, gibt es volumetrisch arbeitende rotierende Kompressoren, die als Zellenverdichter bekannt sind und die heute zur Verbesserung ihres Wirkungsgrades mit Innenkühlung durch Einspritzung für größere Liefermengen ausgeführt werden. Die bekannteste und am weitesten eingeführte Bauart ist die des Kolbenkompressors, die den Vorteil des einfachen Aufbaus und der allgemeinen Kenntnis für sich bucht.

Eine jahrzehntelange Entwicklung an dieser Bauart hat zu

einer sehr betriebssicheren Konstruktion geführt. Kolbenkompressoren werden heute für Druckluftliefermengen von wenigen m³/h bis zu mehreren 1000 m³/h gebaut. Die Druckluftspannung, die man üblicherweise anwendet, ist je nach dem Industriezweig

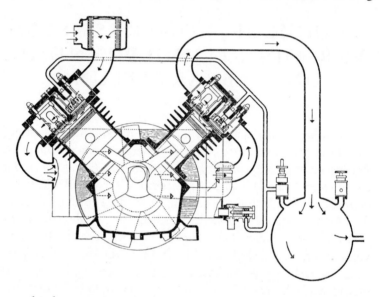

verschieden. Während im Metallgewerbe allgemein 7 atü und darüber verwendet werden, rechnet man im Baugewerbe mit 5 und im Bergbau nur mit 4 atü Druckluftspannung am Werkzeug. Der Kompressor selber hat aber in allen Fällen nahezu denselben Druck zu erzielen, denn die herabgesetzten Druckluftspannungen berücksichtigen nur die in den einzelnen Anwendungszweigen verschiedenen Leitungsverluste.

von F. Frey (Frankfurt am Main)

## VOCABULARY

| | |
|---|---|
| **der Druckluftkompressor** | air compressor |
| **die Steingewinnung** | stone quarrying |
| **die Steinbearbeitung** | stone dressing |
| **die Arbeitsleistung** | work efficiency |
| **das Arbeitsverfahren** | work process |
| **die Forderung** | demand |

| | |
|---|---|
| die Sicherheit | safety |
| das Bedienungspersonal | operating staff |
| die Leistungsfähigkeit | efficiency |
| die Gemeinkosten (pl.) | general expenses |
| die Betrachtung | consideration |
| die Verfügung | disposition, disposal |
| die Drucklufterzeugung | production of compressed air |
| die Fertigung | production |
| das Anwendungsgebiet | field of application |
| die Strömungsmaschine | "flow" machine |
| die Luftmenge | quantity of air |
| der Zellenverdichter | "cell" compressor |
| die Verbesserung | improvement |
| der Wirkungsgrad | operating standard |
| die Innenkühlung | interior cooling |
| die Einspritzung | injection |
| die Liefermenge | delivery |
| der Kolbenkompressor | piston compressor |
| der Vorteil | advantage |
| die Kenntnis | knowledge |
| die Druckluftspannung | air pressure |
| der Industriezweig | branch of industry |
| das Metallgewerbe | metal industry |
| das Baugewerbe | building trade |
| der Bergbau | mining |
| der Fall (Fälle) | case |
| der Druck | pressure |
| der Anwendungszweig | field of application |
| der Leitungsverlust | delivery or transmission loss |

| | | | |
|---|---|---|---|
| ausnützen | utilize | rechnen | calculate |
| anwenden (p.p. angewandt) | apply | berücksichtigen | take into account |
| stellen | set, place | gebräuchlich | usual, customary |
| vorherrschen | predominate | hauptsächlich | chiefly |
| beeinflussen | influence | erhöht | raised, increased |
| steigern | increase | verwunderlich | surprising |
| geschehen | occur, happen | wirtschaftlich | economical |
| antreffen | encounter | maßgebend | decisive(ly) |
| sich beschäftigen mit | occupy oneself with | bestrebt | anxious, endeavouring, intent on |
| ausführen (p.p. ausgeführt) | execute | dadurch | through it, by it (*see Grammar*) |
| buchen | record, register | | |

| | | | |
|---|---|---|---|
| **möglichst** | utmost possible | **jahrzehntelang** | for decades |
| **volumetrisch** | volumetric(ally) | **betriebssicher** | reliable, depend- |
| **rotierend** | rotating | | able |
| **einfach** | simple | **üblicherweise** | usually |

**atü** (abbreviation for **Überdruck-** overpressure, pressure above atmo-
atmosphäre) sphere

**darüber** over that, over it **derselbe** the same (*see*
(*see Grammar*) *Grammar*)
**selber** itself **herabgesetzt** lowered, reduced
**nahezu** almost, practically

*Points for special notice*

**man ist bestrebt, die Arbeitsleistung dadurch zu steigern, daß man
möglichst leistungsfähige Werkzeuge zur Verfügung stellt,** (*Literally*)
one is endeavouring to increase the efficiency *therethrough*, that one places
at the disposal the most efficient tools.

*In idiomatic English:* efforts are being made to increase efficiency by
providing the most efficient tools.

### GRAMMAR
#### COMPOUNDED PREPOSITIONS

We have some of these in English in *thereof, thereon, therefore,
thereupon.*

In German, a preposition followed by a pronoun *which stands
for a thing or things*, e.g. **mit ihm** (*with it*), **vor ihnen** (*in front of
them*), **bei ihr** (*with it*), **an ihm** (*at it*), **durch sie** (*by them*), etc.,
may be replaced by a prepositional compound; for instance, the
above may be expressed by **damit** (*with it*), **davor** (*in front of
them*), **dabei** (*with it*), **daran** (*at it*), **dadurch** (*by them*).

#### THE SAME: DER-, DIE-, DASSELBE

This expression is used as a single word, but both the constituent
parts are declined as though they were still article + adjective.

| | *Singular* | | | *Plural* |
|---|---|---|---|---|
| | M | F | N | M.F.N. |
| N. | **derselbe** | **dieselbe** | **dasselbe** | **dieselben** |
| A. | **denselben** | **dieselbe** | **dasselbe** | **dieselben** |
| G. | **desselben** | **derselben** | **desselben** | **derselben** |
| D. | **demselben** | **derselben** | **demselben** | **denselben** |

# 18

# ELEKTRONENGEHIRNE ALS ÜBERSETZER

Zu Beginn des achtzehnten Jahrhunderts hatte der englische Dichter Jonathan Swift die Idee, daß man alle Weisheit der Welt auf mechanische Weise gewinnen könne. Man brauche nur sämtliche Buchstaben und Wörter in allen möglichen Kombinationen aneinanderzufügen. In „Gullivers Reisen" schreibt er über einen Besuch in der Akademie des utopischen Landes Balnibarbi: „Der Professor befand sich in einem großen Zimmer und war von vierzig Schülern umgeben. Nach der Begrüßung bemerkte er, daß ich ernstlich einen Rahmen betrachtete, welcher den größten Teil des Zimmers in Länge und Breite ausfüllte, und sagte, ich wunderte mich vielleicht, daß er sich mit einem Projekt beschäftige, die spekulativen Wissenschaften durch praktische und mechanische Operationen zu verbessern. Er sei überzeugt, durch seine Erfindung werde die ungebildetste Person bei mäßigen Kosten und nur einiger körperlicher Anstrengung Bücher über Philosophie, Poesie, Staatskunst, Gesetze, Mathematik und Theologie ohne die geringste Hilfe von Geist oder Studium schreiben können."

von KARLHEINZ SCHAUDER

## VOCABULARY

| | | | |
|---|---|---|---|
| das Elektro-<br>nengehirn | electronic brain | das Wort,<br>Wörter | word, words |
| der Übersetzer | translator | die Reise | journey, travel |
| der Beginn | beginning | der Besuch | visit |
| der Dichter | poet, author | das Zimmer | room |
| die Weishelt | wisdom | der Schüler | scholar |
| die Welt | world | die Begrüßung | greeting |
| der Buchstabe | letter | die Länge | length |

73

| | | | |
|---|---|---|---|
| **die Breite** | breadth | **ausfüllen** | fill |
| **das Projekt** | project | **sich wundern** | wonder |
| **die Kosten** | cost, expenditure | **verbessern** | improve |
| (used as a plural) | | **mechanisch** | mechanical |
| **die Anstrengung** | effort | **sämtlich** | all |
| **das Buch** | book | **utopisch** | Utopian |
| **die Staatskunst** | politics | **umgeben** | surrounded |
| **das Gesetz** | law | **ernstlich** | seriously, gravely |
| **die Hilfe** | help, aid | **überzeugt** | convinced |
| **der Geist** | mind, intelligence | (from **überzeugen**) | |
| **gewinnen** | gain, acquire | **ungebildet** | uneducated |
| **aneinanderfügen** | put together, join | **mäßig** | moderate |
| **bemerken** | remark | **körperlich** | physical |

*Points for special note*

**gewinnen könne** might acquire　　　**Er sei überzeugt** He was convinced
**Man brauche** One needed

The verbs here are in the subjunctive mood, which is required in indirect speech, as is explained in the Grammar below. The verbs **wunderte, beschäftige** and **werde** are subjunctive for the same reason.

## GRAMMAR

### THE SUBJUNCTIVE MOOD

In certain circumstances the German verb acquires a different form from that which has appeared, as a rule, in these pages up to this lesson. The forms used have been those of clear, direct statement. Such forms are in the indicative mood.

We are here concerned with the subjunctive mood which has almost disappeared in English. It survives in expressions such as the following (the indicative form is indicated in brackets after each expression).

> *Long* live *the queen!* (*lives*)
> *If only he* were *here!* (*was*)
> Be *he who he may* . . . (*is*)

The subjunctive mood is by no means obsolescent in German. The forms most likely to be met in technical literature are given below.

## Present tense

|       | SEIN  | HABEN | WERDEN | KÖNNEN | GEBEN |
|-------|-------|-------|--------|--------|-------|
| ich   | sei   | habe  | werde  | könne  | gebe  |
| er, sie, |    |       |        |        |       |
| es    | sei   | habe  | werde  | könne  | gebe  |
| wir   | seien | haben | werden | können | geben |
| sie   | seien | haben | werden | können | geben |

## Simple past tense

|       | SEIN  | HABEN  | WERDEN | KÖNNEN  | GEBEN |
|-------|-------|--------|--------|---------|-------|
| ich   | wäre  | hätte  | würde  | könnte  | gäbe  |
| er, sie, |    |        |        |         |       |
| es    | wäre  | hätte  | würde  | könnte  | gäbe  |
| wir   | wären | hätten | würden | könnten | gäben |
| sie   | wären | hätten | würden | könnten | gäben |

Weak verbs have the same form in the simple past tense of the subjunctive as they have in the same tense of the indicative.

Strong verbs and irregular weak verbs modify the stem vowel of the indicative and add **e** and **en** in singular and plural respectively, e.g.

|      | KOMMEN | BRINGEN  |
|------|--------|----------|
| S.   | käme   | brächte  |
| P.   | kämen  | brächten |

*Uses of the Subjunctive*

(*a*) As in English it is used to express a wish.

**Es LEBE die Königin!** *Long* live *the Queen*

(*b*) It is used to express English *would* in the principal clause of a conditional sentence, where a negative is implied.

**Ich WÜRDE es tun, wenn ich es könnte.** *I would do it if I could* ("but I can't" is implied).

**Sie HÄTTE es getan, wenn sie es gekonnt hätte.** *She would have done it if she could have.*

(c) It is used in the "if" clause of a conditional sentence if a negative is implied.

**Wenn es so WÄRE, würden wir nicht arbeiten müssen.**
*If it were so, we should not have to work.*
Which implies: *but it is NOT.*

(d) In indirect statement and question:

**Der Ingenieur bemerkte, daß die Brücke schwach SEI.**
*The engineer remarked that the bridge was weak.*
**Der Detektive fragte den Mann, ob er den Dieb gesehen HABE.** *The detective asked the man if he had seen the thief.*

The use in the reading passage above illustrates this, or an extension of this, use.

**Swift hatte die Idee, daß man alle Weisheit auf mechanische Weise gewinnen KÖNNE.** *Swift had the idea that one could acquire all wisdom by mechanical means.*

(e) *After* the expression meaning *as though,* **als ob,** and *as if,* **als wenn.**

**Der Arbeiter bückte sich, als ob das Gewicht schwer WÄRE.** *The workman bent down as though the weight were heavy.*

(f) In a clause introduced by the conjunction **damit,** *in order that.*

**Der Werkführer sprach laut, damit der Arbeiter sicher VERSTÄNDE.** *The foreman spoke loudly so that the workman would certainly (would be certain to) understand.*

# 19

# DER ERDÖL-KONGRESS, FRANKFURT/MAIN 1963

Es hat einmal eine Zeit gegeben, in der man Lampen mit schmutzigem Öl füllte, das zufällig irgendwo auf der Erde gefunden wurde. . . .

Es hat einmal eine Zeit gegeben, in der man das bei der Leuchtölgewinnung abfallende Benzin als wertloses Nebenprodukt verbrannte. . . .

Es hat einmal eine Zeit gegeben, in der man glaubte, daß die Erdölvorräte der Welt kaum noch länger als 1½ Jahrzehnte reichen würden. . . .

Diese Zeiten sind lange vorbei, obwohl es eine Erdölindustrie erst seit wenig mehr als 100 Jahren gibt.

Heute schätzen Wissenschaftler die bestätigten Erdölreserven auf mehr als 42,2 Milliarden Tonnen. Es ist nach ihrer Ansicht aber auch durchaus wahrscheinlich, daß insgesamt 500 Milliarden Tonnen Erdöl, also mehr als das Zehnfache der Schätzung für Ende 1961, der Erschließung harren. Damit kann der Bedarf selbst bei steigendem Verbrauch noch für Generationen gedeckt werden. Was unternommen werden muß, und wie es möglich sein wird, diese Aufgabe zu bewältigen, soll u. a. auf zahlreichen Fachvorträgen des 6. Welt-Erdöl-Kongresses vom 19–26. Juni 1963 in Frankfurt a. M. erörtert werden.

Das für diesen Zweck von der Zuliefererindustrie den Geologen, Bohrtrupps, Ölgesellschaften, Petrochemikern und Verbrauchern der Produkte zur Verfügung gestellte Zubehör wird während der gleichzeitig auf dem Messe- und Ausstellungsgelände in Frankfurt a. M. stattfindenden Internationalen Ausstellung „inter-oil" von etwa 400 Ausstellern aus aller Welt in beispielloser Vollständigkeit angeboten und von den Besuchern zu besichtigen, zu prüfen, zu vergleichen und zu kaufen sein. Eine einmalige, eine seltene Gelegenheit, die jeder wahrnehmen sollte, wahrnehmen muß, der sich in irgend einer Form mit Erdöl, seiner Erforschung, Suche, Gewinnung, Verarbeitung, Beförderung, mit dem Vertrieb oder dem Verbrauch der Produkte beschäftigt. Allein in 16 Abteilungen der „Informationsschau" wird ein durch instruktive Tafel- und Trickdarstellungen, eindrucksvolle Modellaufbauten, überzeugende Bewegungsabläufe und beispielhafte Musterprodukte der verschiedenen Ver- und Bearbeitungsstadien unterstützter Gesamtüberblick der Erdölindustrie gegeben, wie er in dieser Vollständigkeit kaum jemals wieder gezeigt werden kann.

Es gibt eine Fülle interessanter neuer Instrumente, Geräte, Hilfsmittel, Maschinen und Verfahren, ohne die heute weder die wissenschaftliche noch die industrielle Forschung, weder Geologen, Prospektoren, Erdölgesellschaften und Raffinerien noch Petrochemiker und Verbraucher auskommen können. Sie sind notwendig, um wirtschaftlich und nicht zuletzt wertsteigernd arbeiten zu können, sie muß man kennen, um ihre Möglichkeiten zu überschauen und schon heute zu erproben, was morgen genützt werden muß.

## VOCABULARY

| | | | |
|---|---|---|---|
| **das Erdöl** | petroleum | **die Erschlie-** | opening up, |
| **die Lampe** | lamp | **ßung** | development |
| **die Erde** | earth | **der Bedarf** | need, demand |
| **die Leuchtöl-** | the making of | **der Verbrauch** | consumption |
| **gewinnung** | lamp oils | **die Aufgabe** | task |
| **das Neben-** | by-product | **der Fachvortrag** | lecture |
| **produkt** | | **die Zuliefer-** | supply-industry |
| **der Erdölvorrat** | petroleum reserve | **industrie** | |
| **(-vorräte)** | | **der Bohrtrupp** | drilling team |
| **das Jahrzehnt** | decade | **(-s)** | |
| **der Wissen-** | scientist | **die Ölgesell-** | oil company |
| **schaftler** | | **schaft** | |
| **die Ansicht** | view, opinion | **der Petroche-** | petroleum chemist |
| **das Zehnfache** | ten times | **miker** | |
| **die Schätzung** | estimate | **der Verbraucher** | consumer |

| | | | |
|---|---|---|---|
| das Zubehör | accessories | die Erforschung | research |
| die Messe | fair | die Suche | search |
| die Ausstellung | exhibition | die Gewinnung | extraction |
| das Gelände | land, ground | die Verarbei- | processing |
| der Aussteller | exhibitor | tung | |
| die Vollständig- | completeness, per- | die Beförderung | transport |
| keit | fection | der Vertrieb | sale, distribution |
| der Besucher | visitor | die Abteilung | section |
| die Gelegenheit | opportunity | | |

| | |
|---|---|
| die Informationsschau | information show |
| die Tafeldarstellung | diagrammatic representation, diagram |
| die Trickdarstellung | trick representation, cartoon |
| der Modellaufbau (-ten) | model |
| der Bewegungsablauf | working model |
| das Musterprodukt | sample product |
| das Verarbeitungstadium | manufacturing stage |
| das Bearbeitungstadium | treatment stage |
| der Gesamtüberblick | overall view |

| | | | |
|---|---|---|---|
| die Fülle | abundance | kaufen | buy |
| das Gerät | apparatus | wahrnehmen | make use of |
| das Verfahren | procedure | auskommen | manage |
| die Forschung | research | kennen | know |
| | | überschauen | survey |
| füllen | fill | erproben | test, try |
| gefunden (p.p. | found | nützen | utilize |
| of finden) | | | |
| verbrennen (p.t. | burn up | zufällig | by chance |
| verbrannte) | | irgendwo | somewhere or |
| glauben | believe | | other |
| reichen | reach, last | abfallend | waste, unprofit- |
| schätzen | estimate | | able |
| harren (*followed* | await | wertlos | worthless |
| *by genitive*) | | kaum | scarcely |
| steigen | climb, increase | vorbei | past |
| decken | cover | obwohl | although |
| unternehmen | undertake | bestätigt (from | confirmed |
| (p.p. unter- | | *verb* bestätigen) | |
| nommen) | | durchaus | entirely |
| bewältigen | fulfil | wahrscheinlich | probable |
| erörtern | discuss | insgesamt | in all |
| angeboten (p.p. | offered | selbst | even |
| of anbieten) | | u. a. (unter | among other |
| besichtigen | view, inspect | anderen) | things, *inter alia* |
| vergleichen | compare | stattfindend | taking place |

| | | | |
|---|---|---|---|
| **beispiellos** | unexampled | **beispielhaft** | typical |
| **einmalig** | once-occurring, unique | **jemals** | ever |
| | | **zuletzt** | finally |
| **selten** | rare | **wertsteigernd** | with the object of increasing the value |
| **irgend ein** | some (form) or other | | |
| **eindrucksvoll** | impressive | | |

## Special Points

| | |
|---|---|
| **schätzen . . . auf** | estimate at |
| **nach ihrer Ansicht** | in their opinion |
| **der Erschließung harren** | await the development |
| **bei steigendem Verbrauch** | with increasing consumption |
| **das . . . zur Verfügung gestellte Zubehör** | the accessories put at the disposal |

# 20

# NEUE WEGE BEI DER ERRICHTUNG VON DRUCKLUFTANLAGEN

In den letzten Jahren ist infolge der immer intensiveren Maßnahmen zur Rationalisierung der Fertigung und der ständig fortschreitenden Automation sowohl in der chemischen Industrie wie auch in Maschinenfabriken und Gießereien eine zunehmende Verbreitung der Anwendung von Druckluft festzustellen. Die Druckluft bietet eben für viele Verwendungszwecke so große Vorteile, daß ihr Hauptnachteil, die relativ teure Erzeugung, durch die erzielten Kosteneinsparungen in den Anwendungsgebieten mehr als aufgehoben wird. Vor allem in der Steuerungs und Schalttechnik moderner Arbeitsmaschinen, bei Spannzylindern für Vorrichtungen und Werkzeugmaschinen bis zum vielfältigen Einsatz in explosionsgefährdeten Betrieben hat sich der Anwendungsbereich von Druckluft so vergrößert, wie man es vor wenigen Jahren in Deutschland in diesem Ausmaß noch als unwahrscheinlich angesehen hat. Das gleiche gilt für die steigende Verwendung von Druckluftmotoren, z. B. bei Transfer Anlagen sowie als Drehmomenteneinheiten in den Fließbändern der Automobilindustrie. Überall haben die unbestreitbaren Vorteile: keinerlei Reparaturen über lange Zeit, keinerlei Wartung, keine Wärmeentwicklung, geringes Gewicht, keinerlei Gefahren für das Personal durch elektrischen Strom, der Druckluft Eingang und steigende Verbreitung verschafft.

Die Reinhaltung der Druckluft von Wasserkondensat und Öl, die eine unangenehme Emulsion eingehen und eine völlige Verschmutzung des gesamten Druckluftnetzes zur Folge haben können, ist allerdings Voraussetzung für das einwandfreie Funktionieren der Druckluftgeräte; man sollte sich daher nicht scheuen, die Druckluftnetze ständig zu überprüfen und gegebenenfalls in Ordnung bringen zu lassen. Schon der nachträgliche Einbau von Nachkühlern hinter den Kompressoren kann entscheidende Besserung bringen. Während einer vor einigen Jahren durchgeführten Reise durch die englische Maschinen- und Automobilindustrie konnte der Verfasser feststellen, daß die in diesen Werken verwendete Druckluft von einer Reinheit und Trockenheit war, wie man sie leider bei uns in Deutschland nur in wenigen Fällen findet.

(von Dipl.-Ing. G. P. ALPERS, Essen. Herausgeber:
Demag-Aktiengesellschaft, Pokorny)

## VOCABULARY

| | | | |
|---|---|---|---|
| die Maßnahme | measure, step | die Gießerei | foundry |
| die Rationalisierung | rationalization | die Verbreitung | spread, extension |
| | | die Druckluft | compressed air |
| der Verwendungszweck | | purpose, use | |
| der Hauptnachteil | | chief disadvantage | |
| die Erzeugung | | production | |
| die Kosteneinsparung | | economy | |
| die Steuerungstechnik | | technique of control | |
| die Schalttechnik | | switching technique | |
| der Spannzylinder | | sleeve chuck | |
| die Werkzeugmaschine | | machine tool | |
| der Einsatz | | installation | |
| der Anwendungsbereich | | sphere of application | |
| der Ausmaß | | extent, dimensions | |
| die Verwendung | | use, application | |
| der Druckluftmotor | | motor for compressing air | |
| die Anlage | | factory buildings and plant | |
| die Drehmomenteneinheit | | torque unit | |
| das Fließband | | assembly line | |
| die Reparatur | | repair | |
| die Wartung | | maintenance | |
| die Wärmeentwicklung | | development of heat | |
| das Gewicht | | weight | |
| der Strom | | current | |
| der Eingang | | entry, introduction | |
| die Reinhaltung | | keeping clean | |
| das Wasserkondensat | | condensed water | |

| | |
|---|---|
| die Verschmutzung | dirtying, pollution |
| das Druckluftnetz | compressed air network |
| die Folge | consequence |
| die Voraussetzung | condition, assumption |
| das Funktionieren | functioning |
| das Druckluftgerät | compressed air equipment |

| | | | | |
|---|---|---|---|---|
| die Ordnung | order | | zunehmend | increasing |
| der Einbau | installation, addition | | teuer | dear, expensive |
| | | | erzielt | aimed at, desired |
| der Nachkühler | after-cooler | | vielfältig | manifold |
| die Besserung | improvement | | explosionsge-fährdet | liable to explosion |
| der Verfasser | author | | | |
| die Reinheit | purity | | unwahrschein-lich | improbable |
| die Trockenheit | dryness | | | |
| der Herausgeber | publisher | | überall | everywhere |
| | | | unbestreitbar | indisputable |
| feststellen | establish | | keinerlei | no sort of |
| ist festzustellen | is obvious | | unangenehm | unpleasant |
| bieten | offer | | völlig | complete |
| aufheben (p.p. aufgehoben) | cancel | | gesamt | total |
| | | | allerdings | to be sure |
| vergrößern | increase | | einwandfrei | unobjectionable, flawless |
| ansehen | look on | | | |
| verschaffen | procure, provide | | daher | therefore |
| eingehen | go into | | gegebenenfalls | if occasion arises |
| scheuen | avoid | | nachträglich | subsequent |
| überprüfen | test | | entscheidend | considerable, decided |
| durchführen | carry out | | | |
| infolge | as a result of | | durchgeführt (v. durchführen) | carried out |
| intensiv | intensive | | | |
| sowohl . . . wie auch | not only . . . but also | | verwendet | employed |
| | | | leider | unfortunately |

*For special note*

vor allem, above all

vor wenigen Jahren, a few years ago

das gleiche gilt für, the same applies to

überall haben die unbestreitbaren Vorteile . . . der Druckluft
Eingang und steigende Verbreitung verschafft, the incontestable
advantages have everywhere procured the introduction and increasing
use (spread) of compressed air.

# 21

# DIE DAMPFTURBINE

Der Äolusball des Heron von Alexandrien (um 62 nach Chr.) ist die erste Form einer Dampfturbine. 1600 Jahre später versetzte der Italiener Giovanni de Branca ein Schaufelrad zum Drehen von Bratspießen durch den Dampfstrahl in Bewegung. Die wirklich praktisch brauchbare Anwendung der Dampfturbine zur Erzielung größerer Leistungen gelang 1883 dem Schweden de Laval mit einer einstufigen Gleichdruckturbine. Ein Jahr später folgte C. A. Parsons mit einer Überdruckturbine.

Der Aufbau der Turbine sieht im wesentlichen so aus: In einem geschlossenen Gehäuse ist eine Reihe von feststehenden Laufkränzen mit langen und schmalen Schaufeln angeordnet. Zwischen ihnen liegen die auf der Turbinenwelle angebrachten, ebenfalls mit Stahlschaufeln versehenen Laufräder. Der Dampf strömt durch Düsen ins Innere der Turbine und trifft dort mit großer Geschwindigkeit auf die Schaufeln der Laufräder auf. Diese weichen zufolge ihrer Schräglage dem Dampfdruck aus und setzen die Welle in Bewegung. Die festen Leitkränze leiten den Dampfstrom um und geben ihm die ursprüngliche Strömungsrichtung wieder, so daß er erneut auf anschließende Schaufelkränze wirken kann. Ähnlich wie bei der Kolbenmaschine wird auch hier die Energie des Dampfes stufenweise im Hoch-, Mittel- und Niederdruckteil ausgenützt. Da die Dampfturbine nicht umsteuerbar ist, muß, wo erforderlich, eine besondere Rückwärtsturbine zugefügt werden, wie etwa bei Schiffen. Gerade hier hat sich die Dampfturbine ein weites Anwendungsgebiet geschaffen, vermag doch keine andere Antriebsmaschine derart hohe Leistungen bei verhältnismäßig bescheidenen Platzverhältnissen abzugeben. Dazu kommt noch ihre große Laufruhe.

(Aus „Große Erfindungen," Verlag Hallweg, Bern,
Schweiz)

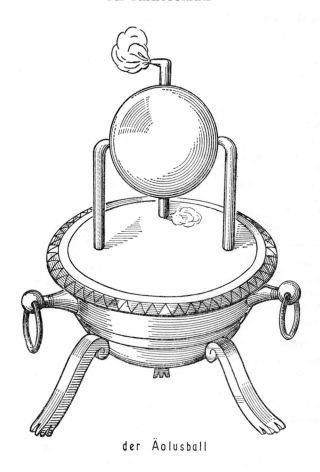

der Äolusball

## VOCABULARY

| | | | |
|---|---|---|---|
| **die Dampftur-<br>bine** | steam turbine | **die Gleichdruck-<br>turbine** | impulse turbine |
| **der Äolusball** | aeolipile or ball of<br>Aeolus | **die Überdruck-<br>turbine** | high-pressure tur-<br>bine |
| **der Italiener** | Italian | **das Gehäuse** | housing, casing |
| **das Schaufelrad** | paddle wheel | **der Laufkranz** | rotating blade |
| **der Bratspieß** | roasting spit | | ring, wheel |
| **der Dampfstrahl** | jet of steam | **die Schaufel** | blade ("shovel") |
| **die Erzielung** | attaining | **die Turbinen-<br>welle** | turbine shaft |
| **der Schwede** | Swede | | |

| | | | |
|---|---|---|---|
| **die Stahlschaufel** | steel blade | **die Geschwin-** | speed, velocity |
| **das Laufrad** | rotor | **digkeit** | |
| **der Dampf** | steam | **der Dampfdruck** | steam pressure |
| **die Düse** | nozzle | **der Leitkranz** | set of redirecting |
| **das Innere** | inside | | fixed blades |
| | | **der Dampfstrom** | current of steam |

| | |
|---|---|
| **die Strömungsrichtung** | current direction |
| **der Schaufelkranz** | ring of blades |
| **die Kolbenmaschine** | piston engine |
| **Hoch-, Mittel- und Niederdruck-** | high, medium and low pressure sec- |
| **teil** | tion(s) |
| **die Rückwärtsturbine** | astern turbine |
| **das Schiff** | ship |
| **die Antriebsmaschine** | engine, motor |
| **das Platzverhältnis** | space condition |
| **die Laufruhe** | smoothness of running |

| | |
|---|---|
| **versetzen** | set, put |
| **folgen** | follow |
| **anordnen** (p.p. **angeordnet**) | dispose |
| **anbringen** (p.p. **angebracht**) | place, fit |
| **versehen** | provide |

Here **versehen** is the past participle: provided

| | |
|---|---|
| **auftreffen (trifft auf)** | impinge on |
| **ausweichen (weicht aus)** | yield, give way |
| **setzen** | set, put |
| **leiten . . . (um)** | lead (round), deflect |
| **wiedergeben** | give back |
| **zufügen** | add |
| **(sich) schaffen** | to make for oneself, itself |
| **vermögen (vermag)** | to be able |
| **abgeben** | deliver, produce |

| | | | |
|---|---|---|---|
| **wirklich** | real(ly) | **anschließend** | adjoining |
| **einstufig** | single stage | **ähnlich** | similar(ly) |
| **geschlossen** | closed | **stufenweise** | in stages |
| (p.p. of **schließen**) | | **umsteuerbar** | reversible |
| **feststehend** | fixed, immobile | **erforderlich** | necessary |
| **schmal** | thin, narrow | **gerade** | just, exactly |
| **ebenfalls** | also, likewise | **derart** | of the kind, similar |
| **zufolge** (+ *geni-* | in consequence | **verhältnismäßig** | comparatively |
| *tive*) | (of) | **bescheiden** | modest |
| **fest** | fixed | **dazu** | in addition, |
| **erneut** | anew, once more | | furthermore |

# 22

# PUNKTE, DIE MAN BEI DER ANSCHAFFUNG EINES WAGENS GANZ BESONDERS BEACHTEN SOLLTE

### 1. *Sichere Straßenlage*

Gute Fahreigenschaften sind das Wichtigste. Davon hängt die Sicherheit der Insassen ab. Wie ein Brett liegt der Maico 500 auf der Straße. Die Gewichtsverteilung ist so vorgenommen, daß der Schwerpunkt sehr tief liegt.

### 2. *Gefedert wie ein großer Wagen*

Alle 4 Räder sind unabhängig voneinander abgefedert: Vorne durch Gummitorsionselemente; die Schwingachsen mit wartungsfreien Gelenken hinten, durch Schraubenfedern. Außerdem hat jedes Rad einen hydraulischen Stoßdämpfer. Auch bei großer Belastung hängt der Wagen hinten niemals durch.

### 3. *Moderne Zahnstangenlenkung*

Die Lenkung eines Autos muß so präzise gearbeitet sein, daß man sich unbedingt auf sie verlassen kann. Die bewährte Zahnstangenlenkung des Maico 500 ist verschleißfest. Beobachten Sie einmal, wie ein Maico-Fahrer sich im Großstadtverkehr bewegt. Durch das günstige Übersetzungsverhältnis der Lenkung und durch ihre Leichtgängigkeit ist der Wagen so wendig, daß der Fahrer auch im dichtesten Verkehrsgewühl eine Lücke findet, durch die er sich hindurchschlängeln kann.

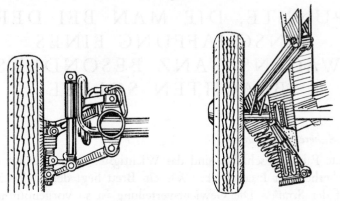

Gefedert wie ein großer Wagen

## 4. *Solide gebautes Chassis*

Der Zentralrohrrahmen mit den Querträgern für die Karosserieaufhängung ist außerordentlich formstabil.

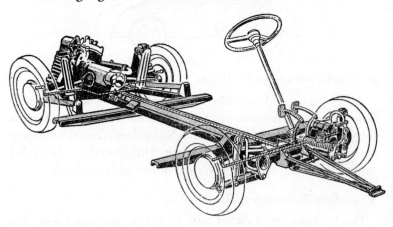

## 5. *Hydraulische Bremsen*

Wichtig sind die Bremsen. Mit den hydraulischen 4-Radbremsen kann der Wagen jederzeit sicher abgestoppt werden.

## 6. *Sparsamer, leistungsstarker Motor*

Wenn man den Fußgashebel durchdrückt, zeigt die Tachonadel auch bei vollbesetztem Wagen bald 90 km/h. Die schweren Fernlaster können dem Maico-Fahrer also nicht imponieren. Er fühlt sich auch beim Überholen dieser Ungetüme völlig sicher. Der wassergekühlte 452-ccm-2-Zylinder-2-Taktmotor hat eine Leistung von 18 PS und ist absolut vollgasfest. Ein Zweitakter hat keine Ventile. Die Konstruktion ist denkbar einfach. Es gibt nur 3 bewegte Teile: Kolben, Pleuel und Kurbelwelle. Daher nur wenig Verschleiß und lange Lebensdauer.

## 7. *4–Gang–Getriebe*

Keine Paßstraße ist zu steil. Für jede Geschwindigkeit und für jede Steigung steht der richtige Gang zur Verfügung. Das Schalten ist ein Kinderspiel.

(Auszüge aus dem Texte eines Prospekts des „Maico 500," eines deutschen Kleinwagens)

## VOCABULARY

| | | | |
|---|---|---|---|
| der Punkt | point, item | das Gummitor-sionselement | rubber torsion element |
| die Anschaffung | purchase, acquisition | die Schwing-achse | independent axle |
| die Straßenlage | road-holding quality | das Gelenk | link, hinge |
| die Fahreigen-schaft | travelling quality | die Schrauben-feder | spiral spring |
| der Insasse (-n) | occupant | die Lenkung | steering |
| das Brett | board | die Zahnstange | rack |
| die Gewichts-verteilung | weight distribution | der Fahrer | driver |
| der Schwer-punkt | centre of gravity | der Großstadt-verkehr | city traffic |

| | |
|---|---|
| das Übersetzungsverhältnis | transmission-ratio |
| die Leichtgängigkeit | ease of movement |
| das Verkehrsgewühl | traffic jam |
| die Lücke | gap |
| der Rohrrahmen | tubular framework |
| der Querträger | crosspiece, transverse support |
| die Karosserieaufhängung | support of the body-work |

| | | | |
|---|---|---|---|
| die Bremse | brake | unabhängig | independent |
| der Fußgashebel | accelerator pedal | wartungsfrei | not requiring maintenance |
| die Tachonadel | speedometer needle | außerdem | besides |
| der Fernlaster | long-distance lorry | niemals | never |
| das Überholen | overtaking | präzise | precisely, accurately |
| das Ungetüm | monster | unbedingt | absolutely |
| das Ventil | valve | bewährt | reliable |
| der Pleuel | connecting rod | verschleißfest | hard wearing, durable |
| die Kurbelwelle | crankshaft | | |
| der Verschleiß | wear and tear | günstig | favourable |
| die Lebensdauer | life | wendig | easily steered, manageable |
| die Paßstraße | mountain pass | | |
| die Steigung | gradient | dicht | thick, dense |
| das Schalten | gear-changing | solide | solidly |
| das Kinderspiel | child's play | außerordentlich | extraordinarily |
| | | formstabil | rigid |
| durchhängen | sag | jederzeit | at all times |
| sich verlassen auf | rely upon | abgestoppt | pulled up |
| | | sparsam | economical |
| beobachten | observe | leistungsstark | strong in performance |
| hindurch- schlängeln (sich) | snake (oneself) through | | |
| | | vollbesetzt | when filled |
| durchdrücken | press down | wassergekühlt | water-cooled |
| imponieren | impress | vollgasfest | steady at full throttle |
| sich fühlen | feel | | |
| | | denkbar | (here) surprisingly |
| sicher | safe | bewegte (Teile) | moving (parts) |
| vorgenommen | arranged, planned | zu | too |
| tief | deep(ly), low | steil | steep |

Special points

**auch bei großer Belastung,** even when heavily loaded.

**bei vollbesetztem Wagen,** when every seat in the car is filled.

**der richtige Gang steht zur Verfügung,** the correct gear is available.

# 23

# FELIX WANKELS
# REVOLUTIONÄRER MOTOR
# (FORTSETZUNG)

Die Vorstellung von einem Motor, dessen Grundbewegung rotierend ist und der doch wie ein Hubkolbenmotor „atmet" und gezündet wird, ist verlockend, gibt aber manche harte Nuß zu knacken auf. Diese Überlegungen haben manchen guten Konstrukteur abgeschreckt. Um so größere Aufmerksamkeit erregte 1959 die Nachricht, die NSU-Werke hätten einen Kreiskolben-Verbrennungsmotor entwickelt, einen sehr kleinen und leichten Motor, der keine Hubkolben, keine Ventile, keine Pleuelstangen, also keinerlei hin- und hergehende Teile und auch keine Kurbelwelle habe und verhältnismäßig billig und einfach herzustellen sei. Wankel hat das Problem auf eine äußerst sinnreiche Weise gelöst. In einem trommelförmigen Gehäuse befindet sich eine Kammer, deren Form etwa einem oben und unten leicht zusammengedrückten Oval entspricht. Darin führt eine dicke Scheibe von der Form eines gleichseitigen Bogendreiecks eine eigenartige Bewegung aus. Dieser Drehkolben rotiert exzentrisch in einer Weise, daß seine drei Ecken ständig den Innenkonturen der Kammer folgen. Er läßt bei dieser Drehbewegung die rundum freiwerdenden Räume abwechselnd kleiner und größer werden und das darin befindliche Gasgemisch sich dementsprechend wechselweise verdichten und ausdehnen. Der Drehkolben bekommt bei jeder Umdrehung drei Schübe, die eine äußerst gleichmäßige Rotation bewirken.

*(Berechtigter Auszug aus der Monatsschrift DAS BESTE AUS READER'S DIGEST, Oktober, 1960)*

Kreiskolbenmotor
*(Querschnittzeichnung)*

① Gas-Luft-Gemisch wird angesaugt

Verdichtet und gezündet ②

③ das verbrannte Gas wird vollständig ausgeschoben

Drei Arbeitstakte des Läufers, während er sich um die Triebwelle dreht

Ansaugkanal

Auslaßkanal

Dichtungen

Triebwelle

Zündkerze

Läufer oder Kolben

## VOCABULARY

| | |
|---|---|
| die Fortsetzung | continuation |
| die Grundbe-wegung | fundamental movement |
| die Nuß | nut |
| die Überlegung | consideration |
| die Aufmerk-samkeit | notice, attention |
| die Nachricht | news |
| der Kreiskolben-motor | rotating piston motor |
| der Hubkolben | piston |
| die Kammer | chamber |
| das Bogendrei-eck | triangle with arched sides |
| der Drehkolben | rotating piston |
| die Ecke | corner |
| die Innenkontur | inside contour |
| die Drehbewe-gung | turning move-ment |
| der Raum | space |
| das Gasgemisch | fuel mixture |
| die Umdrehung | rotation |
| der Schub (ːe) | thrust |
| atmen | breathe |
| knacken | crack |

| | |
|---|---|
| abschrecken | frighten, deter |
| entsprechen | correspond |
| verdichten | compress |
| bewirken | bring about, effect |
| verlockend | alluring, attractive |
| hart | hard |
| mancher | many a |
| hin und her | to and fro |
| äußerst | extremely |
| sinnreich | ingenious |
| trommelförmig | drum-shaped |
| zusammenge-drückt | pressed together |
| gleichseitig | equilateral |
| eigenartig | curious |
| exzentrisch | eccentrically |
| rundum | round about |
| freiwerdend | becoming free |
| abwechselnd | alternately |
| befindlich | to be found |
| dementspre-chend | accordingly |
| wechselweise | alternately |
| gleichmäßig | regular, uniform |
| berechtigt | authorized |

*Special points*

auf eine Weise, in a way

UM SO GRÖßERE Aufmerksamkeit erregte die Nachricht, the news aroused *all the more* attention.

# 24

# ELEKTRONENGEHIRNE ALS ÜBERSETZER 2

Die Geschichte der Kybernetik—und damit auch die der Übersetzungsroboter—ist eng mit der Entwicklung von Rechenmaschinen verbunden. Im siebzehnten Jahrhundert erdachten Pascal und Leibniz die ersten brauchbaren Rechenautomaten. Bereits 1804 wurden immer wiederkehrende Arbeitsvorgänge an Webstühlen mit Lochstreifen gesteuert. 1823 versuchte der Engländer Charles Babbage, eine analytische Rechenanlage zu bauen. Gegen Ende des neunzehnten Jahrhunderts erfand Hollerith die Lochkarte, eines der wesentlichsten Hilfsmittel der Kybernetik. Die automatische Verarbeitung von Daten nahm einen gewaltigen Aufschwung, als vor etwa zwanzig Jahren Konrad Zuse in Deutschland und Howard Aiken in den USA unabhängig voneinander die ersten programmgesteuerten Rechenautomaten konstruierten. Diese Rechenmaschinen bestanden aus einer sinnreichen Verbindung der Elektronik mit dem Dualsystem. Man erkannte die Bedeutung der Erfindung sofort; in schneller Folge entstand eine Unzahl von Modellen, die mit der Zeit immer vollkommener und leistungsfähiger wurden. Heute kann man bereits sagen, daß die Steuerungs- und Informationstheorie die geistige Arbeit in einem ähnlichen Ausmaß umgestalten wird, wie dies durch die Industrialisierung mit der körperlichen Arbeit geschah. Als einer der bedeutendsten Vertreter der Kybernetik in unseren Tagen gilt der amerikanische Professor Wiener.

(von KARLHEINZ SCHAUDER, in ,,Hochland,'' Oktober, 1962)

## VOCABULARY

| | |
|---|---|
| die Kybernetik | cybernetics |
| die Steuerungslehre | the study of control mechanisms |
| der Übersetzungsroboter | translating-robot |
| der Rechenautomat (-en) | computer |
| der Arbeitsvorgang (-gänge) | work-process |
| der Webstuhl (¨e) | loom |
| der Lochstreifen | punched strip |
| die Rechenanlage | calculating machine |
| die Lochkarte | punched card |
| die Daten (*n. pl.*) | data |
| der Aufschwung | impetus |
| die Verbindung | combination |
| die Folge | sequence |
| die Unzahl, eine Unzahl von | endless number, innumerable |
| die Steuerung | control |
| der Vertreter | representative |
| | |
| erdenken (p.t. erdachte) | think out, invent |
| steuern | control, regulate |
| versuchen | try |
| erkennen (p.t. erkannte) | recognize |
| entstehen (p.t. entstand) | arise |
| umstalten (p.p. umgestaltet) | transform |
| geschehen (p.t. geschah) | happen |
| gelten (3rd s.pr.t. gilt) | to be esteemed |

| | | | |
|---|---|---|---|
| eng | narrow(ly), close(ly) | sofort | immediately |
| | | vollkommen | perfect |
| wiederkehrend | recurring | geistig | mental, brain- |
| gegen | towards, against | bedeutend | significant |
| gewaltig | immense (power-ful) | | |

# 25

# ELEKTRONENGEHIRNE ALS ÜBERSETZER 3

Als erster machte sich der Amerikaner Booth daran, eine Art mechanisches Wörterbuch zu konstruieren. Entsprechend der Arbeitsweise der elektronischen Kalkulatoren übersetzte er die Buchstaben und Wörter in Zahlenwerte. Die so umgewandelten Wörter einer Sprache wurden mit den gleichen Ausdrücken einer anderen in das Speicherwerk einer elektronischen Rechenmaschine eingegeben. Die Aufgabe des Kalkulators bestand nun darin, zu einem eingegebenen Zahlenwert einen bestimmten anderen zu suchen, der das gleichlautende Wort in der Fremdsprache darstellte. Zum Beispiel wurde dem Wort „und" die Zahl 314 zugeordnet. Sobald die Maschine diese Zahl von einer Lochkarte abtastete, warf sie die Zahl 524 aus, die in Buchstaben zurückverwandelt die englische Bezeichnung für „und" ergab, also „and". Derartige mechanische Wörterbücher übersetzten jedoch nur Wort für Wort; sie nahmen keine Rücksicht auf grammatikalische oder gar semantische Regeln. Für einen eingegebenen Begriff der einen Sprache warfen sie oft drei oder vier Begriffe der anderen aus, weil dem betreffenden Wort in der Fremdsprache drei oder vier verschiedene Bedeutungen zukamen.

(von KARLHEINZ SCHAUDER in „Hochland," Oktober, 1962)

## VOCABULARY

| | | | |
|---|---|---|---|
| **das Wörterbuch** | dictionary | **die Fremd-** | foreign language |
| **die Arbeitsweise** | mode of operation | **sprache** | |
| **der Zahlenwert** | numerical value | **die Zahl** | number |
| **der Ausdruck** | expression | **die Bezeichnung** | term |
| **das Speicher-** | storage mechan- | **die Rücksicht** | respect |
| **werk** | ism, "memory" | **die Regel (-n)** | rule |

| | | | |
|---|---|---|---|
| **übersetzen** | translate | **bestimmt** | definite, particular |
| **eingeben** | put in, feed into | **gleichlautend** | having the same |
| **bestehen (in)** | consist (in) | | meaning |
| (p.t. **bestand**) | | **sobald** | as soon as |
| **zuordnen** | co-ordinate, link | **zurückverwan-** | changed back |
| **abtasten** | feel out | **delt** | |
| **auswerfen** | throw out | **derartig** | of this sort |
| **ergeben** | yield, supply | **jedoch** | however |
| **zukommen** | fit, apply | **gar** | even, at all |
| | | **semantisch** | semantic |
| **umgewandelt** | transformed | **betreffend** | in question |

*Points for special notice*

**Booth machte sich daran, ein Wörterbuch zu konstruieren,** Booth
   set himself the task of constructing a dictionary.

**die Aufgabe des Kalkulators bestand darin . . . einen anderen zu
   suchen,** the task of the computer consisted in looking for another.

**auf etwas Rücksicht nehmen,** to have regard to or for something.

# 26

## DIE GRENZEN DES ELEKTRONENGEHIRNS

Ein elektronisches Rechengerät kann nicht mehr leisten als der menschliche Geist. Das leistungsfähigste Elektronengehirn in Europa, die IBM 709, vermag zwar in der Minute 2,5 Millionen Additionen auszuführen und die Ergebnisse logisch miteinander zu verbinden. Im Grunde bleibt es jedoch ein Ja-Nein-Sager, der zur Stummheit verdammt ist oder sinnlose Resultate liefert, wenn er nicht vorher von einem vernünftbegabten Menschen mit einem speziellen Programm versehen wurde. Trotz aller technischen Fortschritte können die elektronischen Gehirne nur in ihrer Struktur, nicht aber in ihrer Funktion mit dem menschlichen Gehirn verglichen werden. Wir vermögen zwar die einzelnen Nervenzellen nachzuahmen, aber wir können nicht die Verbindung der zehn Milliarden Neuronen untereinander herstellen. Wir wissen nichts über das Prinzip, auf Grund dessen sie miteinander korrespondieren und schöpferisch wirksam werden. Es ist zum Beispiel nur der menschlichen Phantasie gegeben, sich Dinge anschaulich vorzustellen, die es in der Wirklichkeit noch gar nicht gibt. Es wird wahrscheinlich niemals möglich sein, diesen vielfältigen und weitverzweigten Organisationsplan zu durchschauen und auf elektronische Verhältnisse zu übertragen. Auch in Zukunft bleibt daher dem schöpferischen Menschen ein Reservat vorbehalten, in dem er uneingeschränkt walten kann. Dieser geistige Bereich beginnt jenseits der routinemäßigen Übersetzungen und Analysen, die ein Elektronengehirn auszuführen vermag. Ein Roboter kann letzten Endes nur nachvollziehen, was der gestaltende Mensch vorausberechnet hat. Die Bewunderung, die uns angesichts der gewaltigen und unheimlichen Rechenmaschinen erfaßt, gilt daher eigentlich den Wissenschaftlern und Gelehrten, die sie ersonnen haben. Sie verstanden es, Materie so sinnvoll anzuordnen, daß sie auf Befehl quantitativ mehr leistet als ein menschliches Gehirn. Die Techniker und Ingenieure haben uns aber auch gelehrt, daß eine Maschine nur das Berechenbare erfassen kann, während es gerade die Würde des Menschen ausmacht, unberechenbar zu sein und das Unberechenbare zu tun.

(aus einem Artikel von K. SCHAUDER in „Hochland," Oktober, 1962)

## VOCABULARY

| | | | | |
|---|---|---|---|---|
| das Rechengerät | computer | vorausberechnen | calculate in advance | |
| das Ergebnis | result | erfassen | grip, grasp | |
| der Ja-Nein-Sager | a "yes" or "no" sayer | ersinnen (p.p. ersonnen) | think out, devise | |
| die Stummheit | dumbness | lehren | teach | |
| das Resultat | result | ausmachen | constitute | |
| der Fortschritt | progress | menschlich | human | |
| das Gehirn | brain | zwar | admittedly, it is true | |
| die Nervenzelle | nerve cell | | | |
| das Neuron | neuron | logisch | logically | |
| das Prinzip | principle | verdammt | condemned | |
| die Wirklichkeit | reality | sinnlos | meaningless | |
| | | vorher | previously | |
| das Reservat | preserve, reserve | vernunftbegabt | gifted with intelligence | |
| der Bereich | field, realm | | | |
| die Bewunderung | admiration | trotz | in spite of | |
| | | schöpferisch | creative(ly) | |
| der Gelehrte | savant | wirksam | effective | |
| der Befehl | command | anschaulich | graphically | |
| das Berechenbare | the calculable | weitverzweigt | widely ramified | |
| die Würde | dignity | uneingeschränkt | unrestrained | |
| | | jenseits | beyond | |
| verbinden | combine | routinemäßig | routine | |
| vergleichen (p.p. verglichen) | compare | angesichts (+ gen.) | in the face of | |
| korrespondieren | correspond | unheimlich | uncanny | |
| sich vorstellen | imagine | eigentlich | properly | |
| durchschauen | see through | sinnvoll | meaningfully | |
| walten | rule, hold sway | unberechenbar | incalculable | |
| nachvollziehen | achieve afterwards | | | |

# 27

# EIN NEUES FLUßKRAFTWERK
# IN ÖSTERREICH

In Österreich entsteht das größte Flußkraftwerk Mitteleuropas, das Donaukraftwerk Aschach. In der Reihe der geplanten, insgesamt 15 Staustufen umfassenden Donau-Kraftwerkskette steht es leistungsmäßig mit Abstand an der Spitze: Sein Jahresarbeitsvermögen wird mit in der Regel 1 640 Mio kWh ab Generatorklemmen errechnet.

Bei einer Hebung des Wasserspiegels um rund 16 m über Mittelwasser wird sich der Rückstau über einen Stromabschnitt von 40,6 km erstrecken. Hierbei werden alle Untiefen und Klippen überstaut, so daß diese bisher für die Schiffahrt schwierigste, weil kurven- und hindernisreichste Strecke der Donau völlig ihren wilden Charakter verlieren und zu einem ständig 2-bahnig befahrbaren Abschnitt mit ruhig fließendem Wasser werden wird.

Das Hauptwerk liegt etwa 2 km oberhalb des Ortes Aschach, ohne jedoch diesen selbst und damit seinen aus dem 17. und 18. Jahrhundert stammenden baulichen Charakter zu berühren. Die am rechten Ufer gelegene Schleusenanlage besteht aus zwei Schleusenkammern von je 24 m Breite und 230 m Nutzlänge. Diese Abmessungen gestatten die Aufnahme eines ganzen Schleppzuges, bestehend aus Schleppschiff und 4 paarweise gekoppelten Kähnen von je 1 200 t Nutzlast. Die große Stauhöhe von 15,66 m (bei Mittelwasser) ergibt je Kammer eine Füllmenge von 90 000 bis 120 000 cbm Wasser. Da diese Wassermengen in wirtschaftlich kurzer Zeit nicht ohne Gefahren aus dem Oberhafen entnommen bzw. in den Unterhafen entleert werden können, sind besondere Füllungs- und Entleerungsbauwerke vorgesehen, über die das Wasser direkt dem Hauptstrom entnommen bzw. wieder zugeleitet wird. Die entsprechenden Zu- resp. Abflußkanäle erstrecken sich über die gesamte Schleusenkammersohle und erlauben eine Füllung (Entleerung) in jeweils 12 bis 13 Min. ohne Wasserstandsbeeinflußung des Oberhafens ($100 \times 250$ m) oder des Unterhafens ($100 \times 230$ m). Die Schleusen können außerdem auch zur Ableitung von Katastrophen-Hochwasser und zwar für ca. 28% der ankommenden Wassermenge herangezogen werden.

An das Schleusenunterhaupt schließt stromseits das Krafthaus mit 4 Maschinensätzen und einer im Trennpfeiler zur Wehranlage untergebrachten Eigenbedarfsmaschinenanlage an. Die 4 Turbinen haben—eine Neuerung im Kraftwerkbau—nicht die gleiche Drehrichtung, sondern sind abwechselnd links- und rechtsdrehend angeordnet. Dadurch ist es möglich, je zwei Hauptmaschinen übersichtlich in einem Bedienungsstand zusammenzufassen. Der Nutzungsfaktor ist äußerst günstig: Die Festlegung der Turbinenwassermenge auf 2 040 $m^3$/s bedeutet eine Ausnützung von 95% der zufließenden Wassermenge der Donau.

Die 156 m lange Wehranlage, bestehend aus 4 Wehrpfeilern und 5 Wehrfeldern von je 24 m lichter Weite, stellt die Verbindung zum linken Ufer her.

Der riesige Stauraum von 83 km Uferlänge erfordert natürlich umfangreiche Baumaßnahmen im gesamten Gebiet, so u.a. die Verlegung bzw. den Neuaufbau zweier Dörfer. Einen ungefähren Überblick über die zu vollbringenden Leistungen geben folgende Kubaturen—

| | | |
|---|---|---|
| Treppel- und Wirtschaftswege | rd. | 40 000 $m^3$ |
| Humus (Auf- und Abtrag) | | 100 000 $m^3$ |
| Zwischenboden (Auf- und Abtrag) | | 300 000 $m^3$ |
| Aufhöhungen | | 2 500 000 $m^3$ |
| Böschungspflaster | | 140 000 $m^3$ |
| Steinwurf | | 450 000 $m^3$ |
| Ufer- und Stützmauern | | 44 000 $m^3$ |

Allein für das Hauptbauwerk und den engeren Baubereich

beziffert sich der Felsaushub auf insgesamt 277 000 cbm, der Betonbedarf auf 1 203 500 cbm.

(Mit Erlaubnis des Herausgebers: DEMAG–Baggerfabrik, Werbeabteilung, Düsseldorf–Benrath)

## VOCABULARY

| | | | |
|---|---|---|---|
| **das Flußkraftwerk** | river power station | **das Hauptwerk** | principal construction |
| **Österreich** | Austria | **der Ort** | village |
| **Mitteleuropa** | central Europe | **das Ufer** | bank |
| **die Donau** | Danube | **die Schleusenanlage** | system of locks |
| **die Staustufe** | barrage level | | |
| **die Kette** | chain | **die Schleusenkammer** | lock basin |
| **der Abstand** | distance | | |
| **das Jahresarbeitsvermögen** | annual work capacity | **die Nutzlänge** | effective length |
| | | **die Abmessung** | measurement |
| **Mio (Millionen)** | millions | **die Aufnahme** | admission, taking |
| **Generatorklemmen** | generator terminals | **der Schleppzug** | train of barges |
| | | **das Schleppschiff** | tug |
| **die Hebung** | raising | | |
| **der Wasserspiegel** | water-level | **der Kahn (Kähne)** | boat, barge |
| **das Mittelwasser** | mean water-level | **die Nutzlast** | maximum load |
| **der Rückstau** | impounded water | **die Stauhöhe** | height of the dammed water |
| **der Stromabschnitt** | section of the river | | |
| | | **die Kammer** | chamber, basin |
| **die Untiefe** | shallow, shoal | **die Füllmenge** | total amount |
| **die Klippe** | rock | **der Oberhafen** | upper harbour, or dock |
| **die Schiffahrt** | navigation | | |
| **die Strecke** | stretch | **der Unterhafen** | lower harbour, or dock |
| **der Abschnitt** | section | | |

**das Füllungs-** ⎫
**das Entleerungs-** ⎭ **bauwerk**

filling ⎫
emptying ⎭ structure

**der Hauptstrom** main stream
**der Zufluß-kanal** filling channel

**der Abfluß-kanal** draining channel

**die Schleusenkammersohle**      bottom of the lock basin
**die Wasserstandsbeeinflußung**      influencing the water level

die Ableitung — diversion
die Wassermenge — amount of water
das Schleusenunterhaupt — sluice outlet
der Maschinensatz — machine set
der Trennpfeiler — dividing pier
die Eigenbedarfsmaschine — special purpose machine
die Neuerung — innovation
die Drehrichtung — direction of rotation
der Bedienungsstand — operating stand
der Nutzungsfaktor — utilization factor
die Festlegung — establishing, fixing
die Ausnützung — utilization
die Wehranlage — dam
der Wehrpfeiler — barrage pier
das Wehrfeld — space between the piers
die Weite — width, extent
der Stauraum — reservoir of impounded water
die Uferlänge — bank length
die Baumaßnahme — building measure (s)
die Verlegung — transfer, removal
der Neuaufbau — rebuilding
das Dorf (Dörfer) — village
der Überblick — survey, review
die Kubatur — cubic quantity
der Treppelweg — towing-path
der Wirtschaftsweg — service road
der Auftrag — filling in
der Abtrag — excavation
der Zwischenboden — subsoil
die Aufhöhung — raising

das Böschungspflaster — paving of the slopes
der Steinwurf — shifting of stone
die Mauer — wall
die Stützmauer — supporting wall
der Felsaushub — amount of rock
der Betonbedarf — concrete requirements

die Erlaubnis — permission
der Herausgeber — publisher
der Bagger — excavator, dredger
die Werbeabteilung — advertising department

planen — plan
errechnen — calculate, estimate
erstrecken sich — stretch, extend
überstauen — cover with water
verlieren — lose
fließen — flow
berühren — affect
gestatten — allow
ergeben (ergibt) — yield, result in
entnehmen (p.p. entnommen) — remove, take away
entleeren — empty
vorsehen — provide
zuleiten — return, lead back
erlauben — permit
heranziehen (p.p. herangezogen) — draw upon
anschließen — join, connect
unterbringen (p.p. untergebracht) — house
zusammenfassen — comprise, include
bedeuten — mean
erfordern — require
vollbringen — achieve
sich beziffern — amount

umfassend — embracing, comprising

| | | | |
|---|---|---|---|
| **leistungsmäßig** | as regards output | **bzw. (beziehungsweise)** | or, respectively |
| **um rund** | at about | | |
| **hierbei** | as a result | **resp. (respektive)** | respectively, or |
| **hindernisreich** | full of obstacles | **jeweils** | every time |
| **wild** | wild | **und zwar** | in fact |
| **2–bahnig** | having two lanes | **stromseits** | on the stream side |
| **befahrbar** | navigable | **übersichtlich** | easily |
| **oberhalb** | above | **licht** | clear |
| **stammend (aus)** | dating (from) | **riesig** | gigantic |
| **baulich** | architectural | **umfangreich** | extensive |
| **von je** | each of | **zweier** | of two |
| **paarweise** | in pairs | **ungefähr** | approximate |
| | | **enger** | more limited |

## Special points

**mit Abstand an der Spitze,** far ahead of the others.

**seinen aus dem 17. und 18. Jahrhundert stammenden baulichen Character,** its architectural features deriving from the seventeenth and eighteenth centuries.

# 28

## HERSTELLUNG EINES 200-t-HINTERSTEVENS AUS STAHLGUß VON MAX IBING, VDG, IN BOCHUM

In den letzten Monaten haben Stahlgießereien des Ruhrgebietes vier Hintersteven bisher nicht erreichter Größe für Supertanker von 78 000 t und 88 000 t abgeliefert. Der Hintersteven ist ein wichtiges Element des Schiffaufbaues; er hält das Ruder und den Antriebspropeller (Bild).

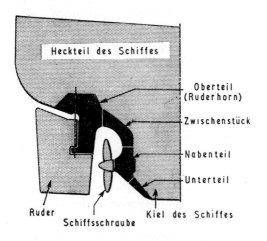

Der hier beschriebene Steven wurde in Rotterdam in einen Tanker eingebaut. Er wiegt 200 t und ist 19 m lang. Die sperrige Konstruktion und die großen Abmessungen erforderten eine Teilung in vier Stücke (Bild). Das größte Einzelstück, das Ruderhorn, wiegt 104 t.

Das für die Formherstellung notwendige Holzmodell wurde in

der eigenen Modellschreinerei des Werkes angefertigt. Zunächst wurde auf dem Fußboden der Modellschreinerei zeichnungsgerecht, jedoch unter Berücksichtigung des Schwindmaßes, ein Aufriß im Maßstab 1 : 1 gezeichnet. Jedes der vier Modellteile wurde in Übereinstimmung mit dem Aufriß getrennt gebaut. Die vier Abgüsse mußten später als Ganzes zusammenpassen. Das Einhalten der vorgeschriebenen Maße erforderte genauestes Anbringen der Kernmarken und Spanten sowie die Verleimung und Absperrung der Verstrebungen und Verbindungen im Modell. An der Herstellung des Modells, das 25 000 kg wiegt, haben acht Modellschreiner insgesamt 4000 Arbeitsstunden gearbeitet. Dabei wurden 80 m³ Holz verbraucht.

Nach Prüfung des Modells wurden die vier Teile einzeln in entsprechend vorbereitete Formgruben eingestampft. Dabei wurde eine besondere feuerbeständige Schamottemasse verwendet, die eine möglichst glatte Oberfläche des Stevens gewährleistete, so daß später keine Angriffsmöglichkeiten für eine zerstörende Kavitation des Wassers vorhanden sind.

Die Gestalt des Ruderhorns mit dem 7,5 m langen Kern erforderte besonders hochwertige handwerkliche Arbeit der Former und Kernmacher. Der 30 t schwere Kern wurde in einem Kernkasten mit einer Spezialkernmasse eingestampft, die sich durch hohe Gasdurchläßigkeit auszeichnet. Zur Versteifung und Sicherung gegen den Auftrieb wurde der Kern mit drei Eisenbahnschienen über die ganze Länge verbaut. Damit der Kern gleichmäßig trocknete, wurde er hängend im Kerntrockenofen gebrannt.

Nach dem Brennen und Nachschwärzen wurde der Kern in die Form eingepaßt. Der hydrostatische Druck des flüssigen Stahls von etwa 600 t machte eine besondere Sicherung der Form erforderlich; sie wurde durch in der Grube verankerte Träger entsprechend gesichert und abgespannt.

Der für die vier Steventeile vorgesehene Stahl wurde im Siemens-Martin-Stahlwerk erschmolzen. Vorgeschrieben war ein unlegierter Kohlenstoffstahl mit maximal 0,23% C, da der Werkstoff gut schweißbar sein muß. Zum Abgießen des Unterteils, des Nabenteils und des Zwischenstückes reichten jeweils

Schmelzen von 60 t aus, während zum Guß des Oberteils 175 t flüssiger Stahl erforderlich waren. Drei Siemens-Martin-Öfen von je 60 t mußten also gleichzeitig beschickt und abgestochen werden. Diese Stahlmenge wurde aus drei Pfannen schnell, aber gleichmäßig durch drei Eingußtrichter in die Form vergossen. Nach 9 min war der Gießvorgang des Oberteils beendet. Je nach Größe wurden einige Trichter nachgegossen oder mit wärmeentwickelnden Lunkermitteln behandelt. Das Trichtergewicht entsprach etwa 75% des Stückgewichts.

Aus der Zeitschrift „Gießerei" 19.4.62

## VOCABULARY

| | |
|---|---|
| die **Herstellung** | construction, manufacture |
| der **Hintersteven** | rear sternpost |
| der **Stahlguß** | cast steel |
| die **Größe** | size |
| der **Schiffaufbau** | ship construction |
| das **Ruder** | rudder |
| das **Bild** | picture, illustration |
| der **Steven** | sternpost |
| die **Teilung** | division |
| das **Stück** | piece, part |
| das **Einzelstück** | single part |
| das **Ruderhorn** | rudder post, rudder-horn |
| die **Formherstellung** | making of the mould, moulding |
| das **Holzmodell** | wooden pattern |
| die **Modellschreinerei** | patternmakers' shop |
| der **Fußboden** | floor |
| die **Berücksichtigung** | regard, consideration |
| das **Schwindmaß** | amount of shrinkage |
| der **Aufriß** | sketch, elevation |
| der **Maßstab** | scale |
| die **Übereinstimmung** | conformity |
| der **Abguß** | casting |
| das **Einhalten** | keeping to, observance |
| das **Anbringen** | bringing together |
| die **Kernmarke** | core-print |
| das **Spant** | rib, frame |
| die **Verleimung** | gluing |
| die **Absperrung** | sealing |
| die **Verstrebung** | bracing, reinforcing, strut |
| der **Modellschreiner** | pattern maker |
| die **Prüfung** | testing |
| die **Formgrube** | moulding pit |
| die **Schamottemasse** | fire-clay mass |
| die **Oberfläche** | surface |
| die **Angriffsmöglichkeit** | possibility of attack |
| die **Kavitation** | pitting |
| die **Gestalt** | form |
| der **Kern** | core |
| der **Former** | moulder |
| der **Kernmacher** | core maker |
| der **Kernkasten** | core box |
| die **Gasdurchlässigkeit** | permeability to gas |
| die **Versteifung** | stiffening |
| die **Sicherung** | insurance, securing |
| der **Auftrieb** | lifting |

| | | | |
|---|---|---|---|
| **die Eisenbahn-schiene** | railway line or rail | **einpassen** | fit into |
| | | **sichern** | secure |
| **der Kern-trockenofen** | core drying oven | **abspannen** (p.p. abgespannt) | loose |
| **das Nach-schwärzen** | blackening | **erschmelzen** (p.p. er-schmolzen) | melt |
| **die Grube** | pit | | |
| **der Träger** | bearer | **ausreichen** | suffice, be sufficient |
| **der Kohlenstoff-stahl** | carbon steel | **beschicken** | charge |
| | | **abstechen** (p.p. abgestochen) | run off, drain |
| **der Werkstoff** | material | | |
| **das Abgießen** | casting | **vergießen** (p.p. vergossen) | pour out |
| **der Nabenteil** | hub piece | | |
| **die Schmelze** | melt | **beenden** | end, finish |
| **die Stahlmenge** | mass of steel | **nachgießen** (p.p. nachgegossen) | add (by pouring) |
| **die Pfanne** | pan | | |
| **der Einguß-trichter** | funnel | **erreicht** | reached |
| | | **beschrieben** | described |
| **der Gießvor-gang** | pouring process | **sperrig** | bulky |
| | | **zeichnungs-gerecht** | in accordance with the drawing |
| **das Lunker-mittel** | feeder compound | | |
| | | **getrennt** | separately |
| **das Trichter-gewicht** | funnel weight | **vorgeschrieben** | prescribed |
| | | **genau** | exact |
| **das Stück-gewicht** | weight of the casting | **entsprechend** | corresponding, suitable |
| **abliefern** | deliver | **vorbereitet** | prepared |
| **einbauen** | build into | **feuerbeständig** | fire resistant |
| **wiegen** | weigh | **glatt** | smooth |
| **zeichnen** | draw | **zerstörend** | destructive |
| **zusammen-passen** | fit together | **vorhanden** | present |
| | | **hochwertig** | high grade |
| **verbrauchen** | use up | **handwerklich** | craftsmanlike |
| **einstampfen** | ram | **hängend** | hanging |
| **gewährleisten** | guarantee | **flüssig** | molten |
| **auszeichnen** (sich) | distinguish | **verankert** | anchored |
| | | **vorgesehen** | intended |
| **verbauen** | build up | **unlegiert** | pure, not alloyed |
| **trocknen** | dry | **schweißbar** | weldable |
| **brennen** (p.p. gebrannt) | burn | **wärmeentwi-ckelnd** | heat-developing, exothermic |

# 29

# AUTOMATION

Automation ist ein Schlagwort unserer Zeit geworden, und es erhebt sich dabei für viele Arbeitnehmer die Frage: „Wird die selbsttätige Maschine den Menschen von seinem Arbeitsplatz verdrängen?"

Muß es so sein, daß Transfer-Bearbeitungsstraßen, Maschinen mit Zubringe-, Meß- und Überwachungseinrichtungen mit elektrischen, pneumatischen und hydraulischen Steuer- und Regelelementen, mit Druckknöpfen, elektrischen Schaltschranken, Anzeige- und Schreibgeräten dem Menschen Angst einjagen? Nein. Den Begriff „Automatisierung" hat C. M. Dolezalek wie folgt definiert—

Automatisierung ist die Befreiung des Menschen von der Ausführung immer wiederkehrender, gleichartiger Verrichtungen und insbesondere eine Loslösung aus der zeitlichen Bindung an den Rhythmus maschineller und anderer technischer Einrichtungen.

Der vernünftige, verständnisbereite Mensch selbst hat es in der Hand, ob diese neue Fertigungstechnik den Menschen wirklich verdrängt oder ob der Mensch ihr Herr bleibt. Dies sollte für die Zukunft unbedingt Mittelpunkt bleiben.

*Voraussetzungen und Mittel für die Automatisierung*

Grundvoraussetzung für alle Überlegungen zum Fertigungsablauf ist eine klare Vorstellung von der Entwicklung des Fertigungsprogramms, d.h. eine Antwort auf die Frage, welche Gußteile in welcher Gestaltung und in welchen Mengen in Zukunft herzustellen sind. Erst wenn diese Fragen eindeutig beantwortet sind, kann man an eine Automatisierung denken.

Oft ist aber nicht einmal der Kunde, der Besteller, in der Lage, eine klare Antwort zu geben, so daß die gesamte Verantwortung beim Druckgußhersteller liegt. Dieser muß daher in den meisten Fällen die notwendigen kalkulatorischen und technischen Konsequenzen für sich ziehen. Um von der technischen Seite her die Kosten so niedrig wie möglich zu halten, ist erforderlich—

(*a*) Automatisierung der Fertigung bei immer wiederkehrendem, gleichem Fertigungsgut in größeren Stückzahlen über einen längeren Zeitraum hinweg;

(*b*) engste Anpassung der Druckgießmaschine in ihrer Bedienung und Leistung an das herzustellende Gußstück;

(*c*) für Entgratung und Bearbeitung Übergang von der magazinlosen zur Magazinmaschine;

(*d*) Anwendung und Verfeinerung von Baukastensystemen, damit Umstellungen auf andere Teile keine zu hohen Kosten verursachen;

(*e*) Konstruktion aller zur Druckgußfertigung notwendigen Maschinen unter dem Blickpunkt der Instandhaltung und Instandsetzung.

## VOCABULARY

| | |
|---|---|
| das Schlagwort | catchword |
| der Arbeitnehmer | employee |
| die Transfer-Bearbeitungsstraße | flow production line |
| die Zubringeeinrichtung | feed device |
| die Meßeinrichtung | measuring device |
| die Überwachungseinrichtung | inspection device |
| der Druckknopf (¨e) | press button |
| der Schaltschrank | switchboard |
| das Anzeige- und Schreibgerät | indicating and recording apparatus |

| | | | |
|---|---|---|---|
| die Angst | fear | die Fertigungstechnik | production technique |
| die Befreiung | liberation | | |
| die Ausführung | carrying out, doing | der Herr | master |
| | | der Mittelpunkt | central point |
| die Verrichtung | job, duty | das Mittel | means |
| die Loslösung | separation, liberation | die Automatisierung | automating |
| die Bindung | obligation, tie | die Grundvoraussetzung | basic condition |
| der Rhythmus | rhythm | | |
| die Einrichtung | arrangement | | |

| | | | |
|---|---|---|---|
| der Fertigungs-ablauf | course of production | die Umstellung | change, conversion |
| die Vorstellung | idea | die Druckguß-fertigung | manufacture of pressure die castings |
| die Antwort | answer | | |
| der Gußteil | casting | | |
| die Gestaltung | design | der Blickpunkt | standpoint, viewpoint |
| die Menge | quantity | | |
| der Kunde | customer | die Instandhaltung | upkeep |
| der Besteller | client | | |
| die Lage | position | die Instandsetzung | repair(ing) |
| die Verantwortung | responsibility | | |
| der Druckguß-hersteller | manufacturer of pressure die castings | sich erheben | raise itself, rise, occur |
| | | verdrängen | oust, displace |
| die Seite | side | einjagen | see "Special points" |
| das Fertigungs-gut | manufactured component | | |
| die Stückzahl | number of pieces | definieren | define |
| der Zeitraum | period of time | beantworten | answer |
| die Anpassung | adjustment | verursachen | cause |
| die Druckgieß-maschine | machine making pressure castings | denken | think |
| die Bedienung | working, operation | wiederkehrend | recurring |
| | | gleichartig | of the same sort |
| das Gußstück | casting | insbesondere | in particular |
| die Entgratung | fettling, de-burring | zeitlich | temporal |
| | | maschinell | mechanical |
| die Bearbeitung | treatment | vernünftig | intelligent |
| der Übergang | transition | verständnis-bereit | willing to understand |
| die Magazin-maschine | machine with a magazine, magazine loading machine | klar | clear |
| | | d.h. (das heißt) | that is |
| | | welch | which |
| | | eindeutig | unambiguously |
| die Verfeine-rung | improvement | meist | most |
| das Baukasten-system | add-a-plant technique | kalkulatorisch | accounting |
| | | magazinlos | without a magazine |

## Special points

**Muß es so sein, daß Transfer-Bearbeitungsstraßen . . . dem Menschen Angst einjagen?** Must it happen that flow production lines inspire fear in man? More idiomatically, "Is it necessary that men should fear flow production lines?"

**immer wiederkehrend,** constantly recurring

**über einen längeren Zeitraum hinweg,** over a considerable period

**unter dem Blickpunkt der Instandhaltung,** bearing in mind the cost of upkeep.

# 30

# B.S.A. TOOLS LTD.
## EINE BERÜHMTE BIRMINGHAMER FIRMA

Die B.S.A. Tools Ltd., eine Filiale der weltbekannten Birmingham Small Arms Co., wurde Ende des letzten Jahrhunderts als eine Spezialfirma für die Herstellung von Spiralbohrern, Reibahlen und Fräsern gegründet. Später wurden auch Sondermaschinen, Vorrichtungen, usw. nach Einzelbestellungen hergestellt.

Die spitzenlose Schleifmaschine der B.S.A. Tools Ltd. wurde 1921 eingeführt, und mit dem ersten B.S.A.-Revolver-Automat begann 1920 eine fortdauernde Reihe von Spitzenleistungen in der Herstellung automatischer Werkzeugmaschinen.

Mit dem ständigen Wachstum der Firma ergab sich die Notwendigkeit einer größeren Betriebsanlage und 1940 wurde der größte Teil der Werkzeugmaschinenbauabteilung der B.S.A. nach der neuen Fabrik bei Kitts Green bei Birmingham überführt. Diese Fabrik erstreckt sich über 10 ha und beschäftigt 1500 der 3000 Mann starken Belegschaft der B.S.A. Tools Ltd. und der Filialbetriebe der B.S.A.-Werkzeugabteilung.

Das Herstellungsprogramm der Firma schließt jetzt sowohl Einspindel- wie auch Mehrspindel-Stangenautomaten und Futterautomaten ein, sowie Revolverdrehbänke, Produktionsdrehbänke, Walzmaschinen für Gewinde und Profile, spitzenlose Schleifmaschinen und Räummaschinen. Werkzeuge und Maschinenzubehörteile werden zu einem ständig wachsenden Ausmaß in der ursprünglichen Fabrik bei Sparkbrook, einem früh industrialisierten Bezirk von Birmingham, weiter hergestellt, während die Räumwerkzeuge und Zubehör der B.S.A. in einer anderen Fabrik bei Redditch in Worcestershire gefertigt werden. Ein bedeutender Grund der ausgedehnten und wirtschaftlichen Produktion der B.S.A. Tools ist die Konstruktion und der Anwendungsbereich der einzigartigen Zusatzeinrichtungen der B.S.A. Maschinen, mit deren Hilfe diese Maschinen für die Herstellung selbst außerordentlich komplizierter Teile in einem Arbeitsgang eingesetzt werden können, wie auch mit Zufuhrmagazingeräten. Zu diesen interessanten Produktionshilfsmitteln gehören auch automatische Kontrollgeräte für die verschiedenen Produktionsstufen sowie das Zusammenketten von Maschinen für die vollständig automatisierte Produktion von Einzelteilen.

Die Büros der Fabrik in Kitts Green umfassen sowohl die Hauptverwaltung der Betriebe und Firmen der B.S.A.-Werkzeugabteilung, wie auch die Abteilungen, die sich mit der Einfuhr von Erzeugnissen ausländischer Herstellungsfirmen befassen, die in Großbritannien durch die Handels-und-Vertriebs-Organisation dieser Abteilung vertreten sind. Die Verkaufskräfte sind in allen wichtigeren Industriebezirken der Britischen Inseln konzentriert, und Verkauf und Vertrieb von B.S.A. Werkzeugmaschinen, Werkzeugen und Zubehör wird von Birmingham aus in Zusammenarbeit mit Büros und Vertretungen in den wichtigsten Ländern der ganzen Welt geleitet.

(from a B.S.A. catalogue, with permission)

## VOCABULARY

| | | | |
|---|---|---|---|
| **die Filiale** | subsidiary | **die Reibahle** | reamer |
| **der Spiralbohrer** | twist drill | **der Fräser** | milling cutter |
| **die Sondermaschine** | | special-purpose machine | |
| **die Einzelbestellung** | | individual requirement | |
| **die Schleifmaschine** | | grinder | |
| **der Revolver-Automat** | | turret screw automatic | |
| **die Spitzenleistung** | | first-class achievement | |
| **das Wachstum** | | growth | |
| **die Notwendigkeit** | | need | |
| **die Betriebsanlage** | | premises | |
| **die Werkzeugmaschinenbauabteilung** | | the section for the construction of machine tools | |
| **ha,** abbreviation for (**das**) **Hektar** | | hectare (2½ acres) | |
| **die Belegschaft** | | staff | |

| | |
|---|---|
| der Filialbetrieb | subsidiary business |
| der Einspindel (Mehrspindel)- Stangenautomat | single-spindle-(multi-spindle) bar automatic |
| der Futterautomat | chucking automatic |
| die Revolverdrehbank | turret lathe |
| die Walzmaschine für Gewinde | thread generator |
| das Profil | section |
| die Räummaschine | broaching machine |
| der Maschinenzubehörteil | machine component |
| der Bezirk | district |
| das Räumwerkzeug | broaching tool |
| die Zusatzeinrichtung | attachment |
| der Arbeitsgang | process |
| das Zufuhrmagazingerät | automatic feed attachment |
| die Produktionsstufe | stage of production |
| das Zusammenketten | linking |
| der Einzelteil | single part |
| die Hauptverwaltung | central administration |
| die Einfuhr | importation |
| das Erzeugnis | product |
| die Handels-Organisation | trading organization |
| die Vertriebs-Organisation | distributing organization |
| die Verkaufskraft | sales force |
| die Vertretung | agency |

| | | | |
|---|---|---|---|
| sich ergeben | arise | berühmt | famous |
| überführen | transfer | spitzenlos | centreless |
| beschäftigen | occupy, employ | fortdauernd | continuing |
| einschließen | include | ausgedehnt | extensive, expand- ed, increased |
| fertigen | produce | | |
| einsetzen | accomplish | einzigartig | unique |
| umfassen | embrace, include | vollständig | completely |
| vertreten | represent | ausländisch | foreign |
| konzentrieren | concentrate | | |

# 31

## RATIONALISIERUNG DER EISENGIEßEREIEN

Um eine weitgehende Rationalisierung der Eisengießereien zu erreichen, wird unter anderem gefordert, daß jederzeit geschmolzenes Eisen mit großer Gleichmäßigkeit der Temperatur und chemischen Zusammensetzung zur Verfügung steht. Schafft man die Möglichkeit, das flüssige Gußeisen in heizbaren Mischern zu sammeln, bevor es vergossen wird, so läßt sich diese Forderung weitgehend erfüllen. Durch Zugabe der verschiedenen Legierungselemente in den Mischer kann man die jeweils gewünschte Zusammensetzung des Eisens erhalten, wobei Schmelz- und Gießbetrieb sich auf einfache und wirtschaftliche Weise aneinander anpassen lassen. Dies kann man durch geeignete Gestaltung des Mischers, z.B. als Vorherd oder größere Einheit, erreichen, wobei die Art des Schmelzofens so gut wie keine Rolle spielt. Die Wirtschaftlichkeit kann ferner durch Wiederverwendung der Schmelzreste aus den Gießpfannen ohne Umschmelzen erhöht werden.

Obwohl die elektrische Energie schon frühzeitig zum Schmelzen von Gußeisen verwendet und dieses Verfahren ständig weiterentwickelt worden ist, trifft dies nicht für die Mischer zu. Allgemein waren sie als gas- oder ölbeheizte Warmhalteöfen ausgebildet. Derartig beheizte Öfen haben jedoch, unter anderem aus metallurgischen Gründen, erhebliche Nachteile. Da der Brenner meist gegen die Badoberfläche gerichtet ist, muß sie zum Erzielen einer guten Wärmeüberführung möglichst frei von Schlacke gehalten werden, falls sich der Ofen nicht ständig in wiegender Bewegung befindet. Die freie Badoberfläche fordert jedoch den unkontrollierbaren Abbrand einer Reihe von Legierungselementen. Zudem wird die Schmelze sehr inhomogen, sowohl bezüglich der Temperatur als auch der chemischen Zusammensetzung, ausgenommen bei mehr oder weniger rotierenden Öfen. Ferner sind die Möglichkeiten zur Überhitzung des Eisens stark begrenzt. Durch die Brennerflammen mit ihrer meist oxydierenden Atmosphäre wird das Ofenfutter stark beansprucht, was oft erhebliche Schwierigkeiten bereitet. Bei Verwendung von Induktionsöfen als Warmhalteöfen werden diese Nachteile vermieden, da die Wärme unmittelbar in der Schmelze erzeugt wird, und außerdem eine metallurgisch günstige Badbewegung entsteht.

## VOCABULARY

| | | | |
|---|---|---|---|
| das Eisen | iron | die Schlacke | slag, dross |
| die Gleich-mäßigkeit | uniformity | der Abbrand | loss in weight |
| | | die Überhitzung | over-heating |
| die Zusammen-setzung | composition | das Ofenfutter | furnace lining |
| | | die Schwierig-keit | difficulty |
| das Gußeisen | cast iron | | |
| der Mischer | mixer | | |
| die Zugabe | addition | fordern | require |
| das Legierungs-element | alloying element | sammeln | collect |
| | | erfüllen | fulfil |
| der Vorherd | preliminary stove | anpassen (sich) | adapt (oneself) |
| der Schmelzofen | blast furnace | zutreffen | be convenient |
| die Rolle | role, part | ausbilden | develop |
| die Wirtschaft-lichkeit | economy, saving | begrenzen | limit |
| | | vermeiden (p.p. vermieden) | avoid |
| die Wiederver-wendung | re-use | | |
| | | wiegen | rock |
| der Schmelzrest | pig metal (that which is left after the melt) | beanspruchen | put to the test, strain |
| | | bereiten | prepare, cause |
| das Umschmel-zen | re-casting | | |
| | | weitgehend | extensive |
| der Warmhalte-ofen | heat-retentive stove | geschmolzen | molten |
| | | heizbar | capable of being heated |
| der Brenner | burner | | |
| die Badober-fläche | surface of the bath | frühzeitig | early |
| | | ölbeheizt | oil-heated |
| die Wärmeüber-führung | heat transfer | erheblich | considerable |
| | | falls | in case |

| | | | |
|---|---|---|---|
| **unkontrollier-bar** | not controllable | **ausgenommen** | except |
| | | **ferner** | moreover |
| **inhomogen** | not homogeneous | **oxydierend** | oxidizing |
| **bezüglich** | relative to | **unmittelbar** | direct |

## Special points

**unter anderem,** among other things.

**falls sich der Ofen nicht ständig in wiegender Bewegung befindet,** in case the furnace is not being rocked continuously.

**so läßt sich diese Forderung erfüllen,** this requirement is fulfilled.

# 32

# TESTPILOTEN AM RANDE
# DES WELTRAUMS

Ehe noch in der kalifornischen Mohavewüste die Sonne aufgeht, arbeitet im kalten Dämmerlicht schon ein Schwarm von Mechanikern auf einem Gerüst unter der rechten Tragfläche eines Großbombers vom Typ B-52. Die Männer hantieren stumm in äußerster Konzentration an der dunklen, rauchenden Nadel eines Flugkörpers, der an einem stromlinienförmigen Aufhängeträger unter der Tragfläche angebracht ist. Der Flugkörper ist eine X-15, die modernste aller Forschungsmaschinen. Sie ist 15 Meter lang, ein Zwerg im Verhältnis zu dem mächtigen Bomber, der sie trägt, und doch ein imponierendes Ding. Man hat mit ihr eine Stundengeschwindigkeit von 6500 Kilometern erreicht und hofft, mit ihr in 80 Kilometer Höhe zu gelangen.

Der Start vom Militärflugplatz Edwards an diesem Morgen soll nur dazu dienen, den Piloten auf einem Flug mit halber Schubleistung in Höhen bis zu 23 Kilometern mit der Maschine vertraut zu machen. Fregattenkapitän Petersen, der neununddreißigjährige Marineflieger, ein hochgewachsener, ruhiger Mensch mit kurzgeschorenem, graugesprenkeltem Haar, ist einer der sechs Piloten, die sich zum Fliegen der X-15 qualifiziert haben. Bei einem Flug mit einer X-15 ist viel zu beobachten und zu tun—und zwar in einem Mindestmaß an Zeit. Der Pilot muß nach dem Sekundenzeiger arbeiten. Auch bei halber Schubleistung verbrennt seine X-15 ihren gesamten Kraftstoff—fast neuneinhalb Tonnen—schon in 115 Sekunden. Wenn alles gut geht, kann er zwischen dem Ausklinkpunkt in 14 000 Meter Höhe und dem Brennschluß in 23 000 Meter Höhe ein Fünfstufenprogramm erledigen. Im Simulator hat er es schon mehr als drei dutzendmal durchexerziert, ist also mit seiner Aufgabe theoretisch bestens vertraut.

Die Testflugreihe, die Petersen an diesem Tag fortsetzt, wurde vor vierzehn Jahren von dem ersten X-Piloten, dem vierundzwanzigjährigen Fliegerhauptmann Yeager, begonnen. Yeager kletterte am 14. Oktober 1947 in die X-1, eine mit Raketentriebwerk ausgestattete Neukonstruktion der Bell Aircraft Company, und konnte mit Vollast kurze Zeit schneller fliegen als der Schall, dessen Geschwindigkeit in Bodennähe 1227 Stundenkilometer beträgt. Das Zeitalter der Überschallgeschwindigkeitsflüge hatte begonnen. Das Durchbrechen der Schallmauer wurde zur Gewohnheit. Neue Generationen von X-Flugzeugen entstanden.

In den letzten vierziger und den ganzen fünfziger Jahren brachten die X-Piloten immer mehr Licht in das Unbekannte. Es gab dabei manche Katastrophe. Ein Pilot büßte sein Leben auf einem Testflug durch die Explosion seiner X-Maschine ein. Schon bald wird man mit der X-15 die äußersten Grenzen des angetriebenen Fluges erreichen. Was dann? ,,Dann,'' sagt Major White, ,,stürzen wir uns auf das Problem der Höchstgeschwindigkeit und Manövrierfähigkeit in niedrigsten Höhen, erforschen, welche aerodynamischen Drücke Flugzeuge noch vertragen und suchen Möglichkeiten, die scheinbaren Grenzen weiter vorzuschieben—wir wissen ja noch nicht die Hälfte von allem.''

(Berechtigter Auszug aus der Monatsschrift *Das Beste aus Reader's Digest*, Januar 1962)

## VOCABULARY

| | | | |
|---|---|---|---|
| der Rand | edge | der Aufhängerträger | carrier |
| der Weltraum | space | | |
| die Wüste | desert | der Zwerg | dwarf |
| das Dämmerlicht | grey light of dawn | die Höhe | height |
| | | der Militärflugplatz | military aerodrome |
| der Schwarm | swarm | | |
| das Gerüst | scaffolding | der Pilot | pilot |
| die Tragfläche | wing | der Flug | flight |
| die Nadel | needle | die Schubleistung | thrust |
| der Flugkörper | flying object | | |

| | | | |
|---|---|---|---|
| die Fregatte | frigate | die Stufe | step, stage |
| der Marine-flieger | naval airman | die Testflug-reihe | series of test flights |
| das Haar | hair | der Hauptmann | captain |
| das Mindestmaß | minimum | das Raketen-triebwerk | rocket motor |
| der Sekunden-zeiger | second hand | die Vollast | full load |
| der Ausklink-punkt | point of release | der Schall | sound |
| | | die Bodennähe | nearness to the ground |
| der Brennschluß | completion of combustion | | |

| | |
|---|---|
| die Stundenkilometer (*plural*) | kilometres per hour |
| das Zeitalter | age |
| die Überschallgeschwindigkeit | ultrasonic speed |
| das Durchbrechen | breaking through |
| die Schallmauer | sound barrier |
| die Gewohnheit | custom |
| das Licht | light |
| das Unbekannte | unknown |
| die Manövrierfähigkeit | manoeuvrability |

| | | | |
|---|---|---|---|
| aufgehen | rise | vertragen | bear, tolerate |
| hantieren | gesticulate | vorschieben | push forward |
| rauchen | smoke | | |
| anbringen (p.p. angebracht) | lodge | ehe | before |
| | | recht | right |
| hoffen | hope | stumm | mutely |
| gelangen | reach, achieve | dunkel | dark |
| dienen | serve | stromlinien-förmig | streamlined |
| tun | do | | |
| erledigen | execute, complete | mächtig | mighty |
| durchexerzieren | practise | doch | yet |
| fortsetzen | continue | imponierend | imposing |
| beginnen (p.p. begonnen) | begin | vertraut | familiar |
| | | hochgewachsen | tall |
| klettern | climb | kurzgeschoren | cut short |
| entstehen (p.p. entstanden) | rise | graugesprenkelt | sprinkled with grey |
| einbüßen (*sep. verb*) | lose, forfeit | dutzendmal | dozen times |
| | | bestens | very well |
| stürzen | fall | ausgestattet | equipped |
| erforschen | explore | angetrieben | propelled |
| | | scheinbar | apparent |

*Special points*

**in einem Mindestmaß an Zeit,** in the minimum of time.

**das Durchbrechen der Schallmauer wurde zur Gewohnheit,** the breaking of the sound barrier became commonplace.

**in den letzten vierziger und den ganzen fünfziger Jahren,** in the late forties and throughout the fifties.

**immer mehr Licht,** more and more light.

# 33

# EIN PAAR B.S.A. MASCHINEN

## SECHSPINDELFUTTERAUTOMAT B.S.A.
### ACME-GRIDLEY, 6″ × 7¼″ BR

Diese Maschine eignet sich für anhaltende Genauigkeit beim schnellen Bearbeiten von Gußstücken, Schmiedestücken und Stahl oder Buntmetallteilen. Jeder Schlitten ist mit separater Nockensteuerung mit unmittelbarer Schlittenbewegung durch die Nocken versehen.

Die Maschine hat vier kräftige Querschlitten und kann auch mit separaten Revolverschlitten versehen werden.

Die Maschine ist mit Gewindeschneideinrichtungen, Hochgeschwindigkeitsbohreinrichtungen und anderen Zusatzeinrichtungen versehen.

| | | |
|---|---|---|
| Nenndurchmesser des Futters Drehdurchmesser (Max) . . . . . | 152 mm | 184 mm |
| Größte Drehlänge . . . . | 177 mm | 203 mm |
| Spindelgeschwindigkeiten (UpM.) . . | 1154–108 | 1045–45 |
| Motor . . . . . . . | 25 PS. | 40 PS. |

### VOCABULARY

| | |
|---|---|
| **die Genauigkeit** accuracy | **das Schmiede-** forging |
| **das Bearbeiten** working, machining | **stück** |
| | **das Buntmetall** non-ferrous metal |
| | **der Schlitten** slide |
| **die Nockensteuerung** | cam control |
| **der Nocken** | cam |
| **der Querschlitten** | cross slide |
| **der Revolverschlitten** | turret slide |
| **die Gewindeschneideinrichtung** | screw cutting, or threading device |

123

| | |
|---|---|
| **die Hochgeschwindigkeits-bohreinrichtung** | high-speed drilling device |
| **der Nenndurchmesser** | nominal diameter |
| **das Futter** | chuck |
| **der Drehdurchmesser** | swing |
| **die Drehlänge** | turning length |
| **UpM (Umdrehungen in der Minute)** | r.p.m. |
| **ein paar** | a few |
| **kräftig** | strong, powerful |

## PRODUKTIONSDREHBÄNKE MIT MEHREREN WERKZEUGEN (152 × 508 MM)

Für das Repetitionsdrehen von Werkstücken zwischen den Spitzen oder im Futter, mit Mehrfachstählen oder einem einzigen, vom Profilgerät gesteuerten Stahl, für Wellen, Spindeln und andere Werkstücke mit mehreren Durchmessern, wenn verschiedene Längs- und Querdreharbeiten erforderlich sind. Die Zusatzeinrichtungen umfassen pneumatische Spannvorrichtungen, Geräte für Winkel- und Profildrehen von vorne, Geräte für gleichzeitiges Drehen von mehreren Durchmessern, sowie Längsdrehanschläge.

| | | | |
|---|---|---|---|
| Drehdurchmesser über den Werkzeugschlitten . | . 152 mm | Swing over tool slides . | 6" |
| | | Swing over bed . . | 12" |
| Werkstücklänge zwischen den Spitzen . . . . | . 540 mm | Admits between centres . | 21¼" |
| Arbeitshub des Supports | . 457 mm | Working stroke of saddle . | 18" |
| Hub der Querschlitten: | | Stroke of cross slides: | |
| vorne | . 82 mm | front . | 3¼" |
| hinten | . 76 mm | rear . | 3" |
| Spindelgeschwindigkeiten | . 63 bis 1400 UpM | Spindle speeds, depending on motor and pulley size. | 63 to 1400 r.p.m. |
| Motor, PS . . . . | . 5 oder 7½ | Motor h.p. . . . | 5 or 7½ h.p. |

## VOCABULARY

| | |
|---|---|
| **das Repetitionsdrehen** | repetition turning |
| **die Spitze** | centre, point |
| **Mehrfachstähle** | multiple tools |
| **das Profilgerät** | former |
| **die Längsdreharbeit** | longitudinal traversing |
| **die Querdreharbeit** | cross traversing |

| die Spannvorrichtung | chuck |
|---|---|
| das Winkeldrehen | taper turning |
| das Profildrehen | form turning |
| der Längsdrehanschlag | length stop |
| einzig | single |

## WAAGERECHTE OBERFLÄCHEN-RÄUMMASCHINE FÜR DAUERBETRIEB

Diese Maschinen sind außerordentlich leicht zu bedienen und brauchen nur vom Bedienungsmann mit Arbeitsstücken versehen zu werden; die Maschine kann auch für automatisches Aufladen eingerichtet werden. Die Arbeitsstücke werden auf Haltevorrichtungen angebracht, die dann automatisch festgehalten und durch endlose Doppelketten unter den Räumnadeln entlang gezogen werden. Die Arbeitsstücke werden in einem Vorgang geräumt und am Abladeende der Maschine abgelegt. Die Maschine eignet sich für flache und unregelmäßige Oberflächen, parallele oder gegeneinander geneigte Oberflächen (im letzteren Fall mit Schaltvorrichtungen). Arbeitsbereich: 15 240 kg. Räumhub: 1680, 2290 oder 3050 mm. Schneidegeschwindigkeiten bis zu 12 500 mm/Min. Modelle für andere Arbeitsbereiche können auch geliefert werden.

### VOCABULARY

| | | | |
|---|---|---|---|
| der Dauerbetrieb | continuous working | die Schneidegeschwindigkeit | cutting speed |
| der Bedienungsmann | operator | | |
| das Aufladen | loading | bedienen | operate |
| die Doppelkette | twin chain | räumen | broach |
| die Räumnadel | broach | ablegen | get rid of, empty |
| das Abladeende | unloading end | neigen | bend, incline |
| die Schaltvorrichtung | indexing fixture | waagerecht | horizontal |
| | | flach | flat, even |
| der Arbeitsbereich | capacity | unregelmäßig | irregular |
| | | gegeneinander | opposite each other |
| der Räumhub | stroke | | |

*Special point*

**gegeneinander geneigte Oberflächen,** surfaces angular to each other.

# 34

# KRAFTWERKE

Ein Kraftwerk ist eine Großanlage zur Umwandlung einer in der Natur vorkommenden Energieart (Primärenergie) in hochwertige elektrische Energie, welche entweder für eigene Fabrikationsanlagen (Industriekraftwerke) oder aber für die Deckung des allgemeinen Energiebedarfs (Industriebetriebe ohne eigene Energieerzeugung, Gewerbe, Haushalt, Verkehr) verwendet wird. Es ist Aufgabe der Kraftwerke, die von den Verbrauchern benötigte Energie zu jedem Zeitpunkt in der vereinbarten Qualität (Spannung und Frequenz) und zu einem möglichst niedrigen Preis zur Verfügung zu stellen.

Zur Zeit stehen folgende *Primärenergien* zur Verfügung—

(a) unmittelbare Energien der Natur, wie Wasser, Wind, Gezeiten, Sonnenenergie und Erdwärme;

(b) chemisch gebundene Energiearten, wie feste Brennstoffe (Kohle, Holz, Torf, Müll), flüssige Brennstoffe (Heizöl, sonstige Kohlenwasserstoffe, Sulfitlauge), gasförmige Brennstoffe (Erdgas, Koksofengas, Gichtgas, sonstige Abgase);

(c) in Atomen gebundene Energie.

Welche der möglichen Energieformen im speziellen Fall einzusetzen ist, hängt von einer Vielzahl von Faktoren ab, zu denen vor allem Preis und Verfügbarkeit der Primärenergie, Struktur des Energiebedarfes (zeitliche Schwankungen), erforderlicher Aufwand für die Umwandlung in elektrische Energie, erreichbarer Wirkungsgrad und erzielbare Betriebssicherheit gehören.

Heute decken die *Wärmekraftwerke*, welche im wesentlichen die chemisch gebundenen Primärenergien verwerten, etwa 75 v. H. des Energiebedarfes der Erde, während der Rest in *Wasserkraftwerken* erzeugt wird. Der Begriff Kraftwerk ist erst durch den

Einsatz von Strömungsmaschinen (Wasser-, Dampf- und Gasturbinen) zur Stromerzeugung und durch die Möglichkeit, die erzeugte Energie über größere Entfernungen zu übertragen, entstanden. Die erste „elektrische Zentrale" wurde noch mit einer Dampfmaschine betrieben und ging im Jahre 1884 mit einer Leistung von 48 kW in Berlin in Betrieb. Im gleichen Jahre wurden in England die ersten Dampfturbinen mit ca. 75 kW Leistung gebaut. Die erste elektrische Energieübertragung fand anläßlich der Weltausstellung von Lauffen am Neckar nach Frankfurt am Main (ca. 170 km) im Jahre 1891 statt.

Aus dem Fischer Lexikon „Technik 3" (*Elektrische Energietechnik*)

## VOCABULARY

| | | | |
|---|---|---|---|
| das Kraftwerk | power station | das Abgas | waste gas |
| die Umwandlung | transformation | die Vielzahl | multiplicity |
| | | die Verfügbarkeit | availability |
| die Natur | Nature | | |
| die Energieart | form of energy | die Schwankung | fluctuation |
| die Deckung | covering, supply | der Aufwand | expenditure |
| das Gewerbe | trade, industry | die Betriebssicherheit | safety of operation |
| der Haushalt | household | | |
| der Verkehr | traffic, communication | das Wärmekraftwerk | thermal power station |
| der Zeitpunkt | moment | die Stromerzeugung | production of current |
| die Frequenz | frequency | | |
| der Preis | price, cost | die Zentrale | power station |
| die Gezeiten | tides | die Übertragung | transmission |
| die Sonnenenergie | energy of the sun | die Weltausstellung | world exhibition |
| die Erdwärme | heat of the earth | | |
| der Torf | peat | verwerten | utilize |
| der Müll | refuse | betreiben (p.p. betrieben) | drive, power |
| das Heizöl | fuel oil | | |
| der Kohlenwasserstoff | hydrocarbon | stattfinden | take place |
| die Sulfitlauge | sulphite liquor | vorkommend | occurring |
| das Erdgas | natural gas | benötigt | required |
| das Koksofengas | coke oven gas | vereinbart | confined |
| das Gichtgas | blast furnace gas | möglichst niedrig | as low as possible |

| | | | |
|---|---|---|---|
| **zur Zeit** | at the moment | **erzielbar** | attainable |
| **gebunden** | combined | **im wesentlichen** | essentially |
| **gasförmig** | gaseous | **in Betrieb gehen** | go into operation |
| **sonstig** | other | **anläßlich** | on the occasion |
| **erreichbar** | attainable | | (of) |

# 35

# TRANSFORMATOREN

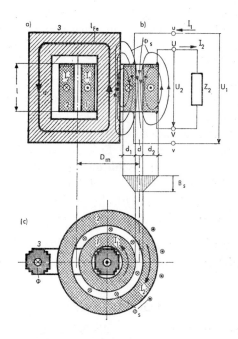

Schnitt (a), Grundriß (c) und Schaltung (b) eines Einphasentransformators. 1 = Innere (Unterspannungs-) Wicklung u–v, Primärwicklung, 2 = äußere (Oberspannungs-) Wicklung U–V, Sekundärwicklung, 3 = Eisenkern, $U_1$ = von außen an die Primärwicklung gelegte Primärspannung, $U_2$ = in der Sekundärwicklung induzierte Sekundärspannung, $I_1$ = Strom der inneren Wicklung (Primärstrom), $I_2$ = Strom der äußeren Wicklung (Sekundärstrom), $Z_2$ = sekundäre Belastungsimpedanz, $\Phi$ = magnetischer Fluß im Eisenkern, $\Phi_s$ = magnetischer Streufluß zwischen den Wicklungen, $B_s$ = größte magnetische Induktion des Streuflusses, $d_1$, $d_2$ = radiale Höhe der Wicklungen, $d$ = radialer Abstand zwischen den Wicklungen, $D_m$ = mittlerer Durchmesser der Wicklungen, $l$ = axiale Länge der Wicklungen, $l_{Fe}$ = mittlere

Länge der Kraftlinien des Flusses $\Phi$ im Eisenkern, $\otimes$ = Pfeil nach hinten, $\odot$ = Pfeil nach vorn.

Transformatoren sind statische Geräte und dienen dazu, elektrische Energie mit einer gegebenen Wechselspannung in solche einer anderen Spannung umzuwandeln. Dies ist notwendig, da elektrische Energie mit Spannungen von 6000 bis 20 000 Volt (6 bis 20 kV) in Kraftwerken erzeugt, mit sehr hohen Spannungen von 50 bis 400 kV auf Freileitungen über große Distanzen übertragen und mit kleinen Spannungen von z. B. 380/220 Volt im Haushalt verbraucht wird. Die Verknüpfung elektrischer Energie verschiedener Spannungen und damit der Bau ausgedehnter Elektrizitätsnetze zur Energieversorgung großer Gebiete ist überhaupt erst durch den Transformator möglich geworden, der 1885 durch K. Zipernowsky, M. Deri und O. Blathy erfunden wurde. In seiner einfachsten Form besteht ein Transformator aus zwei gegeneinander und gegen Erde isolierten *Wicklungen* aus Kupferdraht, welche einen geschlossenen *Eisenkern* umschlingen, wie die Abbildung im Schnitt und im Grundriß zeigt. Der Eisenkern ist geschachtelt aus Blechen hoher Magnetisierbarkeit, den *Transformatorenblechen*, die mit 3–4 v. H. Silizium legiert sind.

Aus dem Fischer Lexikon „Technik 3" (*Elektrische Energietechnik*) 1963. Fischer Bücherei, Frankfurt am Main.

## VOCABULARY

| | | | |
|---|---|---|---|
| **der Transformator** | transformer | **die Belastungsimpedanz** | load impedance |
| **der Schnitt** | section | **der Fluß** | flux |
| **der Grundriß** | outline | **der Streufluß** | eddy current |
| **die Schaltung** | connexion | **die Kraftlinie** | line of force |
| **der Einphasentransformator** | single-phase transformer | **der Pfeil** | arrow |
| **die Unterspannung** | low voltage | **die Wechselspannung** | alternating voltage |
| **die Wicklung** | winding | **die Freileitung** | overhead wires |
| **die Oberspannung** | high voltage | **die Verknüpfung** | linkage, linking |
| **der Eisenkern** | iron core | **das Elektrizitätsnetz** | electricity grid |

| | | | |
|---|---|---|---|
| die Energiever- | supply of energy | legieren | alloy |
| sorgung | | induziert | induced |
| der Kupferdraht | copper wire | mittler | mean, average |
| das Blech | plate | nach hinten | backwards |
| die Magnetisier- | magnetic property | nach vorn | forwards |
| barkeit | | statisch | static |
| das Silizium | silicon | überhaupt | after all |
| | | isoliert | insulated |
| umschlingen | wind round | geschachtelt | packed |

# 36

## DIE STÖRGRÖßEN BEI DER HEIZUNG EINES GEBÄUDES

Zu den Störgrößen zählen wir alle Einflüsse, die auf die Regelgröße, in unserem Fall also auf die Raumtemperatur oder auf die Wärmelieferung, verändernd einwirken. Dazu gehören bei der Raumheizung im wesentlichen die Außentemperatur, die Sonneneinstrahlung, der Windanfall und—was eine Regelung in einer Fabrikhallenheizung unter Umständen stark beeinflussen kann—der Wärmeanfall von Maschinen und anderen Einrichtungen. Die Störgrößen bei industriellen Heizungen stehen vorwiegend mit dem behandelten Produkt und den verwendeten Einrichtungen im Zusammenhang.

Für Raumheizung ist die Außentemperatur die wichtigste Störgröße. Die Anlagen werden allgemein nach der tiefsten Außentemperatur dimensioniert. Nun kommen bekanntlicherweise diese Extremwerte selten vor. Über lange Zeitperioden hin sind die Heizanlagen deshalb nur zum Teil ausgelastet.

Es erübrigt sich hier, auf die Einflüsse der anderen beiden sichtigen und von außen wirkenden Störgrößen bei Raumheizung, auf die Sonneneinstrahlung und den Windanfall, näher einzutreten. Sie stehen ganz in Abhängigkeit der individuellen Konstellation der betreffenden Gebäude und treten immer nur an einem Teil der Gebäudeaußenflächen gleichzeitig auf.

(Aus einem Artikel von W. WIRZ in einem Sonderdruck aus „Technische Rundschau" Bern.)

### VOCABULARY

| | |
|---|---|
| **die Störgröße** | factor of disturbance |
| **die Heizung** | heating |
| **der Einfluß (-flüsse)** | influence |

| die Regelgröße | regulating factor |
| die Raumtemperatur | temperature of the room |
| die Wärmelieferung | supply of heat |
| die Raumheizung | space heating |
| die Außentemperatur | external temperature |
| die Sonneneinstrahlung | sun's rays |
| der Windanfall | onset of wind |
| die Regelung | regulation, adjustment |
| die Fabrikhallenheizung | heating of vast factory space |
| der Umstand | circumstance |
| der Wärmeanfall | onset of warmth |
| der Zusammenhang | connexion |
| die Extremwerte | outside, maximum value |
| die Heizanlage | heating plant |
| die Abhängigkeit | dependence |
| die Konstellation | configuration |
| die Gebäudeaußenfläche | outside surface of building |
| der Sonderdruck | off-print, reprint |
| die Rundschau | review |

| zählen | count |
| einwirken | operate |
| dimensionieren | proportion |
| auftreten | occur |
| verändernd | variably |
| im wesentlichen | in essence, fundamentally |
| unter Umständen | in certain circumstances |
| vorwiegend | predominantly |

| behandelt | dealt with, handled |
| verwendet | employed |
| im Zusammenhang stehen (mit) | be connected with |
| bekanntlicherweise | in a manner well known |
| zum Teil | partly |
| ausgelastet | loaded to capacity |
| sichtig | clear |
| näher | more closely |

*Special point*

**es erübrigt sich** it is unnecessary

# 37

# REGELFRAGEN BEI DEN WÄRMEERZEUGERN

Die Gesamtanschlußwerte der Anlagen, über die hier geschrieben wird, weisen vorwiegend eine Größenordnung von 1–10 Mill. kcal/h auf. In diesen Bereich fallen vor allem in unserem Lande die Leistungswerte der meisten Industrie-Heizungsanlagen. Es handelt sich durchwegs um Hochdruckanlagen mit ständigem Aufsichtspersonal.

In der Praxis kommen sehr viele Varianten von Kesseltypen und Kombinationen von Kesselschaltungen vor, die immer wieder ihre besonderen Anforderungen an die Handhabung stellen. Die Regelung der Wärmeproduktion ist jedoch durchwegs auf sehr einfache Weise gelöst. Die Tendenz zur Automatisierung ist allerdings auch hier fühlbar.

Bei kleinen Kesselanlagen wird vielfach mit einfacher Handsteuerung gearbeitet. Um möglichst gleichbleibende Verhältnisse beim Verfeuerungsprozess zu erhalten, ist es i. allg. vorteilhaft, den Kessel unabhängig von der jeweiligen Belastung des Verteilnetzes und der Verbraucher auf konstante Temperatur oder konstanten Druck zu fahren. Diese Betriebsweise vereinfacht auch die Aufsicht. Der Betrieb mit der höchsten Temperatur kann aber auch von einer ganz anderen Seite her gefordert werden. Wir können damit der Korrosionsgefahr durch Taupunktunterschreitung der Rauchgase im Kessel vorbeugen, was besonders bei der Verbrennung von schweren Heizölen sehr wichtig ist. Dann ist aber die Situation auch meist so, daß auf den einzelnen Verteilsträngen oder dann bei den Verbrauchern noch individuelle und unabhängige Regelungen vorhanden sind.

(Aus einem Artikel von W. Wirz in einem Sonderheft von „Technische Rundschau" Bern.)

## VOCABULARY

| | |
|---|---|
| die Regelfrage | problem of regulation |
| der Wärmeerzeuger | heat-producer |
| der Gesamtanschlußwert | total effective capacity |
| die Größenordnung | order of size |
| der Leistungswert | capacity, effectiveness |
| die Hochdruckanlage | high-pressure plant |
| das Aufsichtspersonal | supervisory staff |
| die Praxis | practice |
| der Kesseltyp | type of boiler |
| die Kesselschaltung | range of boilers |
| die Handhabung | handling, operation |
| die Handsteuerung | hand control |
| der Verfeuerungsprozeß | firing process |
| das Verteilnetz | distribution network |
| die Betriebsweise | mode of operation |
| die Aufsicht | supervision |
| die Korrosionsgefahr | danger of corrosion |
| die Taupunktunterschreitung | control of the dew point |
| das Rauchgas | smoke gas |
| die Verbrennung | burning |
| der Verteilstrang (⁺e) | distribution line |
| das Sonderheft | special number |

| | | | |
|---|---|---|---|
| aufweisen | exhibit | vor allem | above all |
| handeln | concern | durchwegs | without exception |
| stellen (An-forderungen) | make | fühlbar | perceptible |
| | | vielfach | often |
| fahren (Kessel) | work | i. allg. (= im allgemeinen) | in general |
| vereinfachen | simplify | | |
| vorbeugen | prevent | vorteilhaft | advantageous |
| | | jeweilig | existing |

# 38

# REGELUNGSAUFGABEN IN
# DER KLIMATECHNIK

*Einleitung*

Der Begriff der Klimatisierung wurde von den Fachverbänden in den USA und Europa definiert als der „Prozess der Luftbehandlung zum Zwecke der gleichzeitigen Einhaltung von Temperatur, Feuchtigkeit, Reinheit und Verteilung, so daß die Ansprüche des behandelten Raumes erfüllt werden."

Die Klimaanlagen bestehen deshalb aus Luftaufbereitungsanlagen, Verteilsystem für die Luft sowie aus Kälte- und Heizanlagen als Hilfsanlagen.

Die Luftaufbereitung wird je nach Zweck und Unternehmer zentral in einer großen Anlage oder in verschiedene kleinere Anlagen aufgeteilt durchgeführt. Dies kann so weit gehen, daß jeder Fensterachse ein Luftaufbereitungsaggregat zugeordnet ist.

Jedoch sind alle Anlagen Kombinationen von wenigen Elementen wie Filtern, Ventilatoren, Wärmeaustauschern, Drosselklappen und Befeuchtungsvorrichtungen.

*Raumbedingungen*

Die im klimatisierten Raum einzuhaltenden Bedingungen sind je nach Verwendungszweck sehr mannigfaltig. Gewisse industrielle Anlagen verlangen—wie z. B. solche in der Textilindustrie—das Einhalten von sehr engen Grenzen bei der relativen Feuchtigkeit, wobei dann die Temperatur nur einen untern Grenzwert nicht unterschreiten soll. Anderseits wird bei Komfortklimaanlagen eine ziemlich genaue Regulierung der Raumtemperatur gefordert, im Sommer sogar noch in Abhängigkeit der Außentemperatur. Dagegen darf in diesem Fall die relative Feuchtigkeit in ziemlich weiten Grenzen schwanken. Gewisse

Laboratoriumsanlagen verlangen gleichzeitig sehr genaue Einhaltung von Temperatur und Feuchtigkeit.

Um die geforderten Raumbedingungen einhalten zu können, stehen dem Konstrukteur als funktionelle Bestandteile für die Luftaufbereitung nach Temperatur und Feuchtigkeit Mischkammern zur Mischung von Außenluft und Umluft, Heizelemente, Kühlelemente, Befeuchter und Entfeuchter zur Verfügung. Es ist jedoch zu beachten, daß in den meisten Geräten 2 Funktionen

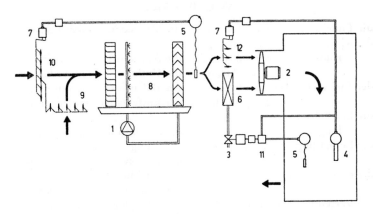

Luftaufbereitungszentrale für Textilklimaanlagen. 1 Pumpe, 2 Ventilator, 3 Heizventil, 4 Hygrostat, 5 Thermostat, 6 Heizregister, 7 Servomotor, 8 Luftwascher, 9 Umluftklappe, 10 Frischluftklappe, 11 Regelverstärker, 12 Bypassluftklappe.

durch physikalische Gesetze gekuppelt sind. So ist z. B. eine Entfeuchtung ohne gleichzeitige Aenderung der Temperatur praktisch nicht zu verwirklichen.

Als 1. Beispiel für die praktische Durchführung der Regelung werde die im Bild dargestellte *Textilklimaanlage* beschrieben. Hier wird die genaue Einhaltung der relativen Feuchtigkeit gefordert. Die Raumtemperatur soll einen untern Grenzwert nicht unterschreiten. Außenluft und Rückluft treten zuerst durch entsprechende Klappen in eine Mischkammer ein und anschließend durch einen Luftwascher, wo praktisch eine Sättigung der Luft mit Wasserdampf stattfindet. Der hinter dem

Wascher angebrachte Temperaturregler steuert durch Veränderung des Verhältnisses von Außenluft resp. Umluftanteil auf konstante Temperatur nach dem Wascher, d.h. konstanten Taupunkt. Diese Anordnung ist deshalb auch als Taupunktregelung bekannt.

(Aus einem Artikel von Dir. K. Sauter, Zürich, in einem
Sonderheft der „Technische Rundschau," Bern, 1959.)

## VOCABULARY

| | | | |
|---|---|---|---|
| die Einleitung | introduction | die Mischkam- | mixing chamber |
| die Klimatisie- | making an artifi- | mer | |
| rung | cial climate, air | die Außenluft | outside air |
| | conditioning | die Umluft | surrounding air |
| der Fachverband | interested party, | der Befeuchter | moistener |
| | group of experts | der Entfeuchter | moisture elimina- |
| die Luftbehand- | air treatment | | tor |
| lung | | die Durchführ- | carrying-out |
| die Einhaltung | retention, mainte- | ung | |
| | nance | der Grenzwert | marginal value |
| die Feuchtigkeit | dampness, | die Rückluft | return air |
| | humidity | die Klappe | valve ("flap") |
| die Verteilung | distribution | die Sättigung | saturation |
| der Anspruch | demand | der Wasser- | water vapour |
| die Klimaanlage | air-conditioning | dampf | |
| | plant | die Veränderung | alteration |
| die Luft | air | der Umluftan- | share of surround- |
| die Luftauf- | air purification | teil | ing air |
| bereitung | | der Taupunkt | dew-point |
| das Verteil- | distribution | die Anordnung | arrangement |
| system | system | der Regelver- | automatic volume |
| die Hilfsanlage | auxiliary plant | stärker | control |
| der Unterneh- | employer | aufteilen | divide up |
| mer | | unterschreiten | go below |
| die Fensterachse | window axis | schwanken | vary, oscillate |
| der Wärmeaus- | heat exchanger | einhalten | maintain |
| tauscher | | kuppeln | link, couple |
| die Drossel- | throttle valve | | |
| klappe | | sowie . . . als | as well . . . as |
| die Befeuchtung | moistening | mannigfaltig | manifold |
| die Raum- | space condition | wobei | by which |
| bedingung | | untern | lower |

| | | | |
|---|---|---|---|
| **anderseits** | on the other hand | **sogar** | even |
| **ziemlich** | fairly | **dagegen** | on the other hand |

## Special points

**Es ist zu beachten,** *it is to be observed*

**eine Entfeuchtung ist . . . praktisch nicht zu verwirklichen,** *the elimination of moisture is not practicable*

Note the passive use of **zu beachten** and **zu verwirklichen**: *to* be *observed, to* be *realized.* Cf. Eng. *to let.*

**Als 1. Beispiel für die praktische Durchführung der Regelung werde die im Bild dargestellte Textilklimaanlage beschrieben,** *As first example of the practical achievement of regulation let the air-conditioning plant (for a textile factory) represented the figure be described.*

Note this use of the subjunctive **werde,** *let be.*

# 39

## LIED DER ARBEIT

Ungezählte Hände sind bereit,
stützen, heben, tragen unsre Zeit.
Jeder Arm, der seinen Amboß schlägt,
ist ein Atlas, der die Erde trägt.

Was da surrt und schnurrt und klirrt und stampft,
aus den Essen glühend loht und dampft,
Räderrasseln und Maschinenklang
ist der Arbeit mächtiger Gesang.

Tausend Räder müssen sausend gehn,
tausend Spindeln sich im Kreise drehn,
Hämmer dröhnend fallen, Schlag um Schlag,
daß die Welt nur erst bestehen mag.

Tausend Schläfen müssen fiebernd glühn,
aber tausend Hirne Funken sprühn,
daß die ewige Flamme sich erhellt,
Licht und Wärme spendend aller Welt.

KARL BRÖGER

### VOCABULARY

| | | | |
|---|---|---|---|
| das Lied | song | das Hirn | brain |
| der Amboß | anvil | der Funke | spark |
| die Esse | chimney, flue | die Wärme | warmth, heat |
| das Räderrasseln | the rattle of wheels | | |
| der Gesang | song | stützen | prop, support |
| der Kreis | circle | heben | heave |
| der Schlag | blow | surren | hum |
| die Schläfe | temple (of head) | schnurren | whirr |

| | | | |
|---|---|---|---|
| **klirren** | clink, clatter | **fiebernd** | feverishly |
| **lohen** | blaze | **sprühen** | spray, sprinkle |
| **dampfen** | steam | **erhellen** | light up |
| **sausen** | rush | **spenden** | bestow |
| **sausend** | rushing(ly) | | |
| **dröhnen** | hum, rumble | **ungezählt** | unnumbered |
| **bestehen** | survive | **bereit** | ready |
| **fiebern** | to be feverish | **ewig** | eternal |

# 40

# DIE STADT ESSEN AN DER RUHR

Umwaldet von den sieben Schornsteinhügeln
der Eisenstadt, wölbt sich ein schwarzer Dom empor
und reißt es auf, das höllenhafte Tor,
aus dem die Eisenreiter, die den Erdball zügeln,

den Atem holen und den Donnerhall der Hufe.
Um seine Mauernflanken kocht die Ruhr, der Rhein,
das Firmament hat keinen Eigenschein,
aus Rad und Dampf schrill klirren Stundenrufe.

Verschollener Väter letzte Enkelscharen
erinnern sich nicht mehr, daß einmal Wiesen waren,
wo jetzt Turbinen donnern und die Eisenbahn.

Im Feuerofen lullt und lockt die Glut
zum Tanz der Schlangen um den Gott Vulkan,
breit klafft ein Spalt: der Erde Schoß speit Blut.

PAUL ZECH

## VOCABULARY

| | | | |
|---|---|---|---|
| **die Stadt** | town | **der Donnerhall** | peal of thunder |
| **der Schornstein-** | hill of chimney- | **der Huf** | hoof |
| **hügel** | stacks | **die Mauern-** | wall-flank |
| **die Eisenstadt** | town of iron | **flanke** | |
| **der Dom** | cathedral | **der Eigenschein** | own radiance |
| **der Eisenreiter** | iron horseman | **der Stundenruf** | call of the hours |
| **der Erdball** | world | **der Vater** | father |
| **der Atem** | breath | **(Väter)** | |

| | | | |
|---|---|---|---|
| **die Enkelschar** | host of grand-children | **holen** | fetch, draw |
| **die Wiese** | meadow | **kochen** | boil |
| **die Eisenbahn** | railway | **sich erinnern** | remember |
| **der Feuerofen** | fiery furnace | **donnern** | thunder |
| **die Glut** | blaze | **lullen** | lull |
| **der Tanz** | dance | **locken** | entice |
| **die Schlange** | serpent | **klaffen** | yawn |
| **der Gott** | god | **speien** | spit |
| **der Spalt** | split | **umwaldet** | surrounded with forest |
| **der Schoß** | bosom | | |
| **das Blut** | blood | **empor** | upwards |
| | | **höllenhaft** | hellish |
| **sich wölben** | arch over | **schrill** | shrilly |
| **reißen (auf)** | tear open | **verschollen** | forgotten, missing |
| **zügeln** | bridle, curb | | |

## Special points

**Verschollener Väter letzte Enkelscharen erinnern sich nicht mehr,**
*The latest brood of now forgotten ancestors (fathers) no longer remembers.*
**der Erde Schoß,** *the bosom of the earth.*

# A SUMMARY OF GERMAN GRAMMAR

## ARTICLES

(A) DEFINITE ARTICLE AND **der** WORDS

|      | Masc. | Fem. | Neut. | Plur. |
|------|-------|------|-------|-------|
| Nom. | der   | die  | das   | die   |
| Acc. | den   | die  | das   | die   |
| Gen. | des   | der  | des   | der   |
| Dat. | dem   | der  | dem   | den   |

**dieser, jeder, welcher,** and **jener** have similar endings.

(B) INDEFINITE ARTICLE, **kein,** AND POSSESSIVE ADJECTIVES

|      | Masc. | Fem. | Neut. | Plural of **kein** and others |
|------|-------|------|-------|-------------------------------|
| Nom. | ein   | eine  | ein   | keine, ihre   |
| Acc. | einen | eine  | ein   | keine, ihre   |
| Gen. | eines | einer | eines | keiner, ihrer |
| Dat. | einem | einer | einem | keinen, ihren |

**mein, dein, sein, unser, euer, ihr,** and **Ihr** have similar endings.

## ADJECTIVES

(A) WITHOUT PRECEDING ARTICLE

|      | Masc. | Fem. | Neut. | Plur. M. F. N. |
|------|-------|------|-------|----------------|
| Nom. | guter Wein   | schöne Kunst  | schweres Blei   | gute Weine   |
| Acc. | guten Wein   | schöne Kunst  | schweres Blei   | gute Weine   |
| Gen. | guten Weins  | schöner Kunst | schweren Bleies | guter Weine  |
| Dat. | gutem Wein   | schöner Kunst | schwerem Blei   | guten Weinen |

## (B) With Definite Article

| | Masc. | Fem. | Neut. | Plur. M.F.N. |
|---|---|---|---|---|
| Nom. | der gute Mann | die gute Frau | das gute Kind | die guten Männer, Frauen, etc. |
| Acc. | den guten Mann | die gute Frau | das gute Kind | die guten Männer, Frauen |
| Gen. | des guten Mannes | der guten Frau | des guten Kindes | der guten Männer, Frauen |
| Dat. | dem guten Mann | der guten Frau | dem guten Kind(e) | den guten Männern, Frauen |

## (C) With Indefinite Article

| | Masc. | Fem. | Neut. |
|---|---|---|---|
| Nom. | ein guter Mann | eine gute Frau | ein gutes Kind |
| Acc. | einen guten Mann | eine gute Frau | ein gutes Kind |
| Gen. | eines guten Mannes | einer guten Frau | eines guten Kindes |
| Dat. | einem guten Mann | einer guten Frau | einem guten Kind(e) |

## COMPARISON OF ADJECTIVES AND ADVERBS

### (A) Adjectives (Regular Comparison)

| Positive | Comparative | Superlative |
|---|---|---|
| klein | kleiner | kleinst |
| praktisch | praktischer | praktischest |

### (Irregular Comparison)

| | | |
|---|---|---|
| gut | besser | best |
| hoch | höher | höchst |
| nah(e) | näher | nächst |
| gern | lieber | liebst |
| viel | mehr | meist |
| groß | größer | größt |

### (B) Adverbs (Regular Comparison)

| Positive | Comparative | Superlative |
|---|---|---|
| klein | kleiner | am kleinsten |
| breit | breiter | am breitesten |
| modern | moderner | am modernsten |

### (Irregular Comparison)

| gut | besser | am besten |
|---|---|---|
| hoch | höher | am höchsten |
| nah(e) | näher | am nächsten |
| gern | lieber | am liebsten |
| viel | mehr | am meisten |
| groß | größer | am größten |

### PERSONAL PRONOUNS

| | | | | | | | | |
|---|---|---|---|---|---|---|---|---|
| *Nom.* | ich | du | er | sie | es | wir | ihr | sie |
| *Acc.* | mich | dich | ihn | sie | es | uns | euch | sie |
| *Gen.* | meiner | deiner | seiner | ihrer | seiner | unser | euer | ihrer |
| *Dat.* | mir | dir | ihm | ihr | ihm | uns | euch | ihnen |

### RELATIVE PRONOUNS

#### (Singular)

| | | | |
|---|---|---|---|
| *Nom.* | ein Mann, **der** . . . | eine Frau, **die** . . . | ein Kind, **das** . . . |
| *Acc.* | ein Mann, **den** . . . | eine Frau, **die** . . . | ein Kind, **das** . . . |
| *Gen.* | ein Mann, **dessen** . . . | eine Frau, **deren** . . . | ein Kind, **dessen** . . . |
| *Dat.* | ein Mann, **dem** . . . | eine Frau, **der** . . . | ein Kind, **dem** . . . |

#### (Plural)

| | |
|---|---|
| *Nom.* | Männer, Frauen, Kinder, **die** . . . |
| *Acc.* | Männer, Frauen, Kinder, **die** . . . |
| *Gen.* | Männer, Frauen, Kinder, **deren** . . . |
| *Dat.* | Männer, Frauen, Kinder, **denen** . . . |

**welcher** is an alternative relative pronoun which is similarly declined.

### INTERROGATIVE PRONOUNS

| | *People* | *Things* |
|---|---|---|
| *Nom.* | **wer?** | **was?** |
| *Acc.* | **wen?** | **was?** |
| *Gen.* | **wessen?** | **wessen?** |
| *Dat.* | **wem?** | — |

## POSSESSIVE ADJECTIVES

| | |
|---|---|
| **mein,** *my* | **unser,** *our* |
| **dein,** *thy* | **euer,** *your* (familiar) |
| **sein,** *his* | **ihr,** *their* |
| **ihr,** *her* | **Ihr,** *your* (formal) |
| **sein,** *its* | |

These are all declined like **ein.**

## PREPOSITIONS

WITH ACCUSATIVE ONLY

**durch, für, gegen, ohne, um, wider.**

WITH DATIVE ONLY

**aus, außer, bei, binnen, entgegen, gegenüber, gemäß, mit, nach, nächst, nebst, samt, seit, von, zu, zuwider.**

WITH GENITIVE ONLY

**anstatt, statt, außerhalb, innerhalb, oberhalb, unterhalb, diesseits, jenseits, halber, inmitten, kraft, längs, laut, trotz, während, wegen, vermöge.**

WITH DATIVE OR ACCUSATIVE (For rules see p. 25)

**an, auf, hinter, in, neben, über, unter, vor, zwischen.**

## CONJUNCTIONS

CO-ORDINATING

**und, aber, sondern, allein** (but), **denn, oder.**

SUBORDINATING

The following are the commonest—

**daß, als, wenn, weil, da, ob, ehe, bevor, obgleich,**

**obwohl, obschon, seitdem, nachdem, damit** (*so that, in order that*).

(For use see p. 42)

## VERBS

### WEAK VERB

| | *Present* | *Imperfect* | *Perfect* | *Future* |
|---|---|---|---|---|
| (ich) | mache | machte | habe gemacht | werde machen |
| (du) | machst | machtest | hast gemacht | wirst machen |
| (er) | macht | machte | hat gemacht | wird machen |
| (wir) | machen | machten | haben gemacht | werden machen |
| (ihr) | macht | machtet | habt gemacht | werdet machen |
| (sie) | machen | machten | haben gemacht | werden machen |

| *Present Subjunctive* | | *Past Subjunctive* |
|---|---|---|
| (ich) | mache | machte |
| (du) | machest | machtest |
| (er) | mache | machte |
| (wir) | machen | machten |
| (ihr) | machet | machtet |
| (sie) | machen | machten |

### STRONG VERB

| | *Present* | *Imperfect* | *Perfect* | *Future* |
|---|---|---|---|---|
| (ich) | gebe | gab | habe gegeben | werde geben |
| (du) | gibst | gabst | hast gegeben | wirst geben |
| (er) | gibt | gab | hat gegeben | wird geben |
| (wir) | geben | gaben | haben gegeben | werden geben |
| (ihr) | gebt | gabt | habt gegeben | werdet geben |
| (sie) | geben | gaben | haben gegeben | werden geben |

*Present Subjunctive*

| (ich) | gebe | (wir) | geben |
|---|---|---|---|
| (du) | gebest | (ihr) | gebet |
| (er) | gebe | (sie) | geben |

*Past Subjunctive*

| | |
|---|---|
| (ich) gäbe | (wir) gäben |
| (du) gäbest | (ihr) gäbet |
| (er) gäbe | (sie) gäben |

## AUXILIARY VERBS

### SEIN, *to be*

| | INDICATIVE | | SUBJUNCTIVE | |
|---|---|---|---|---|
| | *Present* | *Imperfect* | *Present* | *Imperfect* |
| (ich) | bin | war | sei | wäre |
| (du) | bist | warst | seist | wärest |
| (er) | ist | war | sei | wäre |
| (wir) | sind | waren | seien | wären |
| (ihr) | seid | wart | seit | wäret |
| (sie) | sind | waren | seien | wären |

*Future:* ich werde sein, du wirst sein, etc.
*Perfect:* ich bin gewesen, du bist gewesen, etc.

### HABEN, *to have*

| | INDICATIVE | | SUBJUNCTIVE | |
|---|---|---|---|---|
| | *Present* | *Imperfect* | *Present* | *Imperfect* |
| (ich) | habe | hatte | habe | hätte |
| (du) | hast | hattest | habest | hättest |
| (er) | hat | hatte | habe | hätte |
| (wir) | haben | hatten | haben | hätten |
| (ihr) | habt | hattet | habet | hättet |
| (sie) | haben | hatten | haben | hätten |

*Future:* ich werde haben, du wirst haben, etc.
*Perfect:* ich habe gehabt, du hast gehabt, etc.

# WERDEN

|  | INDICATIVE | | SUBJUNCTIVE | |
|--|-----------|-----------|-----------|-----------|
|  | *Present* | *Imperfect* | *Present* | *Imperfect* |
| (ich) | werde | wurde | werde | würde |
| (du) | wirst | wurdest | werdest | würdest |
| (er) | wird | wurde | werde | würde |
| (wir) | werden | wurden | werden | würden |
| (ihr) | werdet | wurdet | werdet | würdet |
| (sie) | werden | wurden | werden | würden |

*Future:*　**ich werde werden,** etc.
*Perfect:*　**ich bin geworden (worden),** etc.

## AUXILIARY VERBS OF MOOD

| | KÖNNEN | MÜSSEN | WOLLEN | SOLLEN | DÜRFEN | MÖGEN |
|--|--------|--------|--------|--------|--------|-------|
| *Present:* | | | | | | |
| ich kann | muß | will | soll | darf | mag | |
| du kannst | mußt | willst | sollst | darfst | magst | |
| Sie können | müssen | wollen | sollen | dürfen | mögen | |
| er kann | muß | will | soll | darf | mag | |
| wir können | müssen | wollen | sollen | dürfen | mögen | |
| ihr könnt | müßt | wollt | sollt | dürft | mögt | |
| Sie können | müssen | wollen | sollen | dürfen | mögen | |
| sie können | müssen | wollen | sollen | dürfen | mögen | |

| | KÖNNEN | MÜSSEN | WOLLEN | SOLLEN | DÜRFEN | MÖGEN |
|--|--------|--------|--------|--------|--------|-------|
| *Imperfect:* | | | | | | |
| ich konnte | mußte | wollte | sollte | durfte | mochte | |
| du konntest | mußtest | wolltest | solltest | durftest | mochtest | |
| Sie konnten | mußten | wollten | sollten | durften | mochten | |
| er konnte | mußte | wollte | sollte | durfte | mochte | |
| wir konnten | mußten | wollten | sollten | durften | mochten | |
| ihr konntet | mußtet | wolltet | solltet | durftet | mochtet | |
| Sie konnten | mußten | wollten | sollten | durften | mochten | |
| sie konnten | mußten | wollten | sollten | durften | mochten | |

*Perfect:*
**ich habe**

| gekonnt | gemußt | gewollt | gesollt | gedurft | gemocht |
|---------|--------|---------|---------|---------|---------|

*Future:* **ich werde können, müssen, wollen, sollen, dürfen, mögen.**

*Present Subjunctive:* **ich könne, müsse, wolle, solle, dürfe, möge.**

*Past Subjunctive:* **könnte, müßte, wollte, sollte, dürfte, möchte.**

## SEPARABLE PREFIXES

**ab-, an-, auf-, aus-, bei-, daher-, dahin-, dar-, ein-, empor-, entgegen-, entzwei-, fort-, her-, hin-** (and their compounds, e.g. **herein, hinaus**), **inne-, los-, mit-, nach-, nieder-, statt-, teil-, vor-, voran-, voraus-, vorbei-, vorüber-, weg-, zu-, zurück-, zusammen-.**

Verbs prefixed by **durch-, hinter-, über-, um-, unter-, voll-, wieder-** may be separable or inseparable.

If the verb with such a prefix can be translated *literally* it is usually separable, e.g. **wiederholen** is separable when it means *fetch back.* It is inseparable when it means *repeat.*

# COMMON ABBREVIATIONS

| | | |
|---|---|---|
| AG. | **Aktiengesellschaft** | joint stock company, Ltd. |
| bezw. | **beziehungsweise** | respectively, or |
| ca. | **circa** | about |
| d.h. | **das heißt** | i.e. |
| d.i. | **das ist** | i.e. |
| Dipl.-Ing. | **Diplom-Ingenieur** | Certificated Engineer |
| DRGM. | **Deutsches Reichs-gebrauchsmuster** | German registered design |
| DRP. | **Deutsches Reichspatent** | German patent |
| F. | **Fahrenheit** | |
| G.m.b.H. | **Gesellschaft mit beschränkter Haftung** | limited liability company |
| ha | **hektar** | |
| Kilo | **Kilogramm** | |
| M or Mk. | **Mark(s)** | |
| N.B. | *nota bene* | N.B. |
| n.Chr. | **nach Christi Geburt** | A.D. |
| p. | *per, pro, pagina* | by, for, page |
| p.a. | *per annum* | |
| PS | **Pferdestärke** | horse-power |
| q | **Quadrat** | square |
| S. | **Seite** | page |
| s. | **siehe** | see |
| t | **Tonne** | ton |
| u. | **und** | and |
| u.a. | **unter anderen** | amongst other things, *inter alia* |
| ü. (u) d.M. | **über (unter) dem Meerespiegel** | above (below) sea level |
| UpM | **Umdrehungen pro Minute** | r.p.m. |

| usf. | **und so fort** | and so on, etc. |
|------|----------------|-----------------|
| usw. | **und so weiter** | and so on, etc. |
| u.zw. | **und zwar** | and indeed (it is often unnecessary to translate this) |
| v. | **von, vom** | |
| v.Chr. | **vor Christi Geburt** | B.C. |
| vorm. | **vormittags** | in the morning, a.m. |
| z.B. | **zum Beispiel** | e.g. |
| z.T. | **zum Teil** | partly |

# LIST OF STRONG AND
# MIXED VERBS

Compounds formed from these verbs are not given unless they are conjugated differently. Less common alternative forms are shown in brackets. The vowel of the Imperfect Subjunctive is not given unless it is irregular. Verbs conjugated with **sein** are marked *. Even these take **haben** when they are transitive.

| Infinitive | Present Indicative (3rd pers. sing.) | Imperfect Indicative (1st pers. sing.) | Past Participle | Imperative | Imp. Subj. | English |
| --- | --- | --- | --- | --- | --- | --- |
| backen | bäckt | buk | gebacken | backe | | bake |
| befehlen | befiehlt | befahl | befohlen | befiehl | ö (ä) | command |
| beginnen | beginnt | begann | begonnen | beginne | ö (ä) | begin |
| beißen | beißt | biß | gebissen | beiß | | bite |
| bergen | birgt | barg | geborgen | birg | | protect, hide |
| *bersten | birst | barst (borst) | geborsten | birst | ö (ä) | burst |
| betrügen | betrügt | betrog | betrogen | betrüge | | deceive |
| biegen | biegt | bog | gebogen | biege | | bend |
| bieten | bietet | bot | geboten | biete | | offer |
| binden | bindet | band | gebunden | binde | | bind |
| bitten | bittet | bat | gebeten | bitte | | ask |
| blasen | bläst | blies | geblasen | blase | | blow |
| *bleiben | bleibt | blieb | geblieben | bleibe | | remain |
| braten | brät | briet | gebraten | brate | | roast |
| brechen | bricht | brach | gebrochen | brich | | break |
| brennen | brennt | brannte | gebrannt | brenne | brennte | burn |
| bringen | bringt | brachte | gebracht | bringe | | bring |
| denken | denkt | dachte | gedacht | denke | | think |
| *dringen | dringt | drang | gedrungen | dringe | | press |
| dürfen | darf | durfte | gedurft | — | | be allowed |
| empfehlen | empfiehlt | empfahl | empfohlen | empfiehl | ö (ä) | recommend |
| *erlöschen | erlischt | erlosch | erloschen | erlisch | | be extinguished |
| *erschrecken | erschrickt | erschrak | erschrocken | erschrick | | be frightened |
| essen | ißt | aß | gegessen | iß | | eat |

| Infinitive | Present Indicative (3rd pers. sing.) | Imperfect Indicative (1st pers. sing.) | Past Participle | Imperative | Imp. Subj. | English |
|---|---|---|---|---|---|---|
| *fahren | fährt | fuhr | gefahren | fahre | | go |
| *fallen | fällt | fiel | gefallen | falle | | fall |
| fangen | fängt | fing | gefangen | fange | | catch |
| finden | findet | fand | gefunden | finde | | find |
| *fliegen | fliegt | flog | geflogen | fliege | | fly |
| fliehen | flieht | floh | geflohen | fliehe | | flee |
| *fließen | fließt | floß | geflossen | fließe | | flow |
| fressen | frißt | fraß | gefressen | friß | | eat (of animals) |
| *frieren | friert | fror | gefroren | friere | | freeze |
| geben | gibt | gab | gegeben | gib | | give |
| *gehen | geht | ging | gegangen | gehe | | go |
| *gelingen | gelingt | gelang | gelungen | es gelinge | | succeed |
| gelten | gilt | galt | gegolten | gilt | ö (ä) | be worth |
| genießen | genießt | genoß | genossen | genieße | | enjoy |
| *geschehen | geschieht | geschah | geschehen | es geschehe | | happen |
| gewinnen | gewinnt | gewann | gewonnen | gewinne | ö (ä) | win |
| gießen | gießt | goß | gegossen | gieße | | pour |
| gleichen | gleicht | glich | geglichen | gleiche | | resemble |
| *gleiten | gleitet | glitt | geglitten | gleite | | glide |
| graben | gräbt | grub | gegraben | grabe | | dig |
| greifen | greift | griff | gegriffen | greife | | grasp |
| haben | hat | hatte | gehabt | habe | | have |
| halten | hält | hielt | gehalten | halte | | hold |
| hangen | hängt | hing | gehangen | hange | | hang |

| Infinitive | Present Indicative (3rd pers. sing.) | Imperfect Indicative (1st pers. sing.) | Past Participle | Imperative | Imp. Subj. | English |
|---|---|---|---|---|---|---|
| heben | hebt | hob | gehoben | hebe | | lift |
| heißen | heißt | hieß | geheißen | heiße | | be called |
| helfen | hilf | half | geholfen | hilf | ü (ä) | help |
| kennen | kennt | kannte | gekannt | kenne | kennte | know |
| klingen | klingt | klang | geklungen | klinge | | sound |
| *kommen | kommt | kam | gekommen | komme | | come |
| können | kann | konnte | gekonnt | — | | be able |
| *kriechen | kriecht | kroch | gekrochen | krieche | | creep |
| laden | lädt | lud | geladen | lade | | load |
| lassen | läßt | ließ | gelassen | laß | | let |
| *laufen | läuft | lief | gelaufen | laufe | | run |
| leiden | leidet | litt | gelitten | leide | | suffer |
| leihen | leiht | lieh | geliehen | leihe | | lend |
| lesen | liest | las | gelesen | lies | | read |
| liegen | liegt | lag | gelegen | liege | | lie |
| meiden | meidet | mied | gemieden | meide | | shun |
| messen | mißt | maß | gemessen | miß | | measure |
| mögen | mag | mochte | gemocht | — | | like |
| müssen | muß | mußte | gemußt | — | | be obliged |
| nehmen | nimmt | nahm | genommen | nimm | | take |
| nennen | nennt | nannte | genannt | nenne | nennte | name |
| pfeifen | pfeift | pfiff | gepfiffen | pfeife | | whistle |
| *quellen | quillt | quoll | gequollen | quill | | gush out |
| raten | rät | riet | geraten | rate | | advise |

157

| Infinitive | Present Indicative (3rd pers. sing.) | Imperfect Indicative (1st pers. sing.) | Past Participle | Imperative | Imp. Subj. | English |
|---|---|---|---|---|---|---|
| reiben | reibt | rieb | gerieben | reibe | | rub |
| reißen | reißt | riß | gerissen | reiße | | tear |
| *reiten | reitet | ritt | geritten | reite | | ride |
| *rennen | rennt | rannte | gerannt | renne | rennte | run |
| riechen | riecht | roch | gerochen | rieche | | smell |
| *rinnen | rinnt | rann | geronnen | rinne | ö (ä) | flow |
| rufen | ruft | rief | gerufen | rufe | | call |
| saugen | saugt | sog | gesogen | sauge | | suck |
| schaffen | schafft | schuf | geschaffen | schaffe | | create |
| *schallen | schallt | scholl | geschollen | schalle | | resound |
| *scheiden | scheidet | schied | geschieden | scheide | | part |
| scheinen | scheint | schien | geschienen | scheine | | shine, seem |
| schieben | schiebt | schob | geschoben | schiebe | | push |
| schießen | schießt | schoß | geschossen | schieße | | shoot |
| schlafen | schläft | schlief | geschlafen | schlafe | | sleep |
| schlagen | schlägt | schlug | geschlagen | schlage | | strike |
| *schleichen | schleicht | schlich | geschlichen | schleiche | | creep |
| schließen | schließt | schloß | geschlossen | schließe | | shut |
| schlingen | schlingt | schlang | geschlungen | schlinge | | sling |
| *schmelzen | schmilzt | schmolz | geschmolzen | schmilz | | melt |
| schneiden | schneidet | schnitt | geschnitten | schneide | | cut |
| schreiben | schreibt | schrieb | geschrieben | schreibe | | write |
| schreien | schreit | schrie | geschrie(e)n | schreie | | scream |
| *schreiten | schreitet | schritt | geschritten | schreite | | stride |

| Infinitive | Present Indicative (3rd pers. sing.) | Imperfect Indicative (1st pers. sing.) | Past Participle | Imperative | Imp. Subj. | English |
|---|---|---|---|---|---|---|
| schweigen | schweigt | schwieg | geschwiegen | schweige | | be silent |
| *schwellen | schwillt | schwoll | geschwollen | schwill | | swell |
| *schwimmen | schwimmt | schwamm | geschwommen | schwimme | ö (ä) | swim |
| schwingen | schwingt | schwang | geschwungen | schwinge | | swing |
| sehen | sieht | sah | gesehen | sieh | | see |
| *sein | ist | war | gewesen | sei | | be |
| senden | sendet | sandte | gesandt | sende | sendete | send |
| singen | singt | sang | gesungen | singe | | sing |
| *sinken | sinkt | sank | gesunken | sinke | | sink |
| sitzen | sitzt | saß | gesessen | sitze | | sit |
| sollen | soll | sollte | gesollt | — | sollte | owe |
| spinnen | spinnt | spann | gesponnen | spinne | ö (ä) | spin |
| sprechen | spricht | sprach | gesprochen | sprich | | speak |
| *springen | springt | sprang | gesprungen | springe | | jump |
| stechen | sticht | stach | gestochen | stich | | sting |
| stehen | steht | stand | gestanden | stehe | | stand |
| stehlen | stiehlt | stahl | gestohlen | stiehl | ä (ü) | steal |
| *steigen | steigt | stieg | gestiegen | steige | | mount |
| *sterben | stirbt | starb | gestorben | stirb | | die |
| stoßen | stößt | stieß | gestoßen | stoße | ü | push |
| streichen | streicht | strich | gestrichen | streiche | | stroke |
| tragen | trägt | trug | getragen | trage | | carry |
| treffen | trifft | traf | getroffen | triff | | meet, hit |
| treiben | treibt | trieb | getrieben | treibe | | drive |

| Infinitive | Present Indicative (3rd pers. sing.) | Imperfect Indicative (1st pers. sing.) | Past Participle | Imperative | Imp. Subj. | English |
|---|---|---|---|---|---|---|
| *treten | tritt | trat | getreten | tritt | | step |
| trinken | trinkt | trank | getrunken | trinke | | drink |
| tun | tut | tat | getan | tue | | do |
| verbergen | verbirgt | verbarg | verborgen | verbirg | | hide |
| *verderben | verdirbt | verdarb | verdorben | verdirb | ü | spoil |
| vergessen | vergißt | vergaß | vergessen | vergiß | | forget |
| verlieren | verliert | verlor | verloren | verliere | | lose |
| verschwinden | verschwindet | verschwand | verschwunden | verschwinde | | disappear |
| verzeihen | verzeiht | verzieh | verziehen | verzeihe | | pardon |
| *wachsen | wächst | wuchs | gewachsen | wachse | | grow |
| waschen | wäscht | wusch | gewaschen | wasche | | wash |
| weben | webt | wob | gewoben | webe | | weave |
| *weichen | weicht | wich | gewichen | weiche | | give away |
| weisen | weist | wies | gewiesen | weise | | show |
| wenden | wendet | wandte | gewandt | wende | wendete | turn |
| werben | wirbt | warb | geworben | wirb | | woo |
| *werden | wird | wurde (ward) | geworden | werde | ü | become |
| werfen | wirft | warf | geworfen | wirf | ü | throw |
| wiegen | wiegt | wog | gewogen | wiege | | weigh |
| winden | windet | wand | gewunden | winde | | wind |
| wissen | weiß | wußte | gewußt | wisse | | know |
| wollen | will | wollte | gewollt | wolle | wollte | wish |
| ziehen | zieht | zog | gezogen | ziehe | | pull, move |
| zwingen | zwingt | zwang | gezwungen | zwinge | | force |

# GERMAN-ENGLISH
# VOCABULARY

Nouns and other words used as nouns are printed with their articles.

In the case of frequently occurring nouns, the plural suffix is given in brackets. If a vowel is modified in the plural, this is indicated by two dots over the hyphen: ⸚.

(irr.) following a verb indicates that it is irregular. Its principal forms may be found in the list of Strong and Mixed verbs on p. 154 *et seq*.

The bracketed numbers after the German words are those of the sections in which each word first appears.

**der Abbrand** (31) (⸚e), loss in weight
**der Abend** (-e), evening
**abends** (5), in the evening
**aber** (1), but
**abfallend** (19), waste, unprofitable
**abfedern** (13), spring
**der Abflußkanal** (27), drainage channel
**das Abgas** (34) (-e), waste gas
**die Abgasturbine** (34), exhaust gas turbine
**abgeben** (21), deliver, produce
**abgestoppt** (22), pulled up
**das Abgießen** (28) (irr.), casting (process)
**der Abguß** (28) (⸚e), casting
**abhängen** (2) (**von**) (irr.), depend (on)
**die Abhängigkeit** (36), dependence
**das Abkühlen**, cooling
**das Abladeende** (33), unloading end
**ablegen** (33), get rid of, empty
**die Ableitung** (27), diversion
**abliefern** (28), deliver

**die Abmessung** (27), measurement
**abschneiden** (4), cut off
**der Abschnitt** (27) (-e), section
**abschrecken** (23), frighten
**die Absicht** (-en), intention
**abspannen** (28), loosen
**die Absperrung** (28), blocking, obstruction
**der Abstand** (27) (⸚e), distance
**abstechen** (28), run off, drain
**abstellen** (6), put, put away
**abtasten** (25), feel out
**die Abteilung** (19) (-en), section
**der Abtrag** (27), excavation
**abwechselnd** (23), alternately
**ägyptisch** (9), Egyptian
**ähnlich** (21), similar
**akkurat** (7), accurate
**alle** (12), all
**allein** (1), alone
**allerdings** (20), of course
**alles** (7), everything
**allgemein** (2), general
**als** (2), when, as, than

**also** (10), so, thus, therefore
**alt** (8), old
**altmodisch** (7), old-fashioned
**der Amboß** (39) (**-e**), anvil
**anbieten** (16) (irr.), offer
**der Anblick** (5), view
**anbringen** (21) (irr.), lodge, place, fit
**das Anbringen** (28), bringing together
**ander** (1), other
**andererseits** (11), on the other hand
**anderseits** (38), on the other hand
**die Änderung** (13), alteration, change
**aneinanderfügen** (18), put together, join
**anfahren** (15) (irr.), start off
**das Anfahren** (15), starting off
**der Anfang** (10), beginning
**anfangs** (9), at first
**anfertigen** (10), prepare
**die Anforderung** (16) (**-en**), demand
**angebracht** (32), fitted, fixed
**angehen** (11) (irr.), concern
**angehören** (6), belong
**angesichts** (26), in the face (of)
**angetrieben** (32), propelled
**die Angriffsmöglichkeit** (28), possibility of attacking
**die Angst** (29) (**-̈e**), fear, anxiety
**anhalten** (14) (irr.), stop, check
**ankommen** (5) (irr.), arrive
**ankommen** (**auf**), depend (on)
**die Anlage** (20) (**-n**), plant, installation
**anläßlich** (34), on the occasion (of)
**anordnen** (21), arrange, dispose
**die Anordnung** (38) (**-en**), arrangement
**anpassen** (31) (**sich**), adapt (oneself)
**die Anpassung** (29) (**-en**), adjustment
**anregend** (11), interesting, captivating (ly)

**die Anregung** (11) (**-en**), stimulation, impulse
**ansaugen** (23) (irr.), suck in
**der Ansaugkanal** (23) (**-̈e**), inlet
**die Anschaffung** (22), purchase, acquisition
**anschaulich** (26), graphically
**anschließen** (27) (irr.), join, connect
**anschließend** (21), finally, adjoining
**ansehen** (20), look on
**die Ansicht** (19) (**-en**), view
**anspringen** (15) (irr.), start up
**der Anspruch** (38) (**-̈e**), demand
**die Anstrengung** (18) (**-en**), effort
**antreffen** (17) (irr.), encounter
**antreiben** (6) (irr.), drive, motivate
**der Antrieb** (12) (**-e**), motive force, power, drive
**die Antriebskraft** (**-̈e**), motive power
**die Antriebsmaschine** (21), engine, motor
**die Antwort** (29), answer
**die Anweisung** (15) (**-en**), instruction
**anwenden** (17), use, apply
**die Anwendung** (12) (**-en**), application
**der Anwendungsbereich** (20), sphere of application
**das Anwendungsgebiet** (17) (**-e**), field of application
**der Anwendungszweig** (17), field of application
**das Anzeigegerät** (29) (**-e**), indicating apparatus
**der Äolusball** (21), aeolipile
**die Arbeit** (2) (**-en**), work
**arbeiten** (1), work
**der Arbeiter** (1), worker
**der Arbeitnehmer** (29), employee
**der Arbeitsanzug** (6) (**-̈e**), overall
**der Arbeitsbereich** (33), capacity
**der Arbeitsgang** (30), process, cycle

die **Arbeitsleistung** (17), work efficiency

der **Arbeitstakt** (23), work stroke

das **Arbeitsverfahren** (17), work process

der **Arbeitsvorgang** (24) (⸚e), work process

die **Arbeitsweise** (25), mode of operation

**arbeitswillig** (6), ready and willing to work

**arm** (5), poor

die **Art** (2) (-en), kind, sort

der **Artikel** (2), article

die **Asche** (5) (-en), ash

der **Atem** (40), breath

**atmen** (23), breathe

**auch** (1), also

**auf** (6), on

der **Aufbau** (12), construction

die **Aufgabe** (19) (-n), task

**aufgeben** (16), give up, set (problem)

**aufgehängt** (13), suspended

**aufgehen** (32) (irr.), rise

**aufgesetzt** (13), applied, pushed on

der **Aufhängerträger** (32), carrier

**aufheben** (15) (irr.), release, cancel

die **Aufhöhung** (27), raising

**aufhören** (6), stop

**aufkommen** (9) (irr.), rise, appear

das **Aufladen** (33), loading

die **Aufmerksamkeit** (23), notice, attention

die **Aufnahme** (27) (-n), admission

der **Aufriß** (28) (-e), sketch, elevation

der **Aufschwung** (24), impetus

das **Aufsehen** (14), stir, sensation

die **Aufsicht** (37), supervision

das **Aufsichtspersonal** (37), supervisory staff

**aufteilen** (38), divide up

der **Auftrag** (27), filling in

**auftreffen** (21) (irr.), impinge on

**auftreten** (36) (irr.), appear, occur

der **Auftrieb** (28), lifting

der **Aufwand** (34), expenditure

**aufweisen** (37) (irr.), exhibit

**aus**, out of, from

**ausbilden** (31), develop, train

**ausbohren** (9), bore

**ausbrennen** (5) (irr.), burn out

**ausdehnen** (**sich**) (14), expand, extend

der **Ausdruck** (25) (⸚e), expression

**ausfliegen** (14) (irr.), fly out

**ausführen** (17), execute

**ausführlich** (12), in detail

die **Ausführung** (29), carrying out, doing

**ausfüllen** (18), fill

**ausgedehnt** (30), extensive

**ausgelastet** (36), loaded to capacity

**ausgenommen** (31), except

**ausgestattet** (32), equipped

der **Ausklinkpunkt** (32), point of release

**auskommen** (19) (irr.), manage

**ausländisch** (30), foreign

der **Auslaßkanal**, outlet (channel)

**auslegen** (16), lay out

**ausmachen** (26), constitute

das **Ausmaß** (20), dimensions, extent, range, scale

**ausnützen** (17), utilize

die **Ausnützung** (27), exploitation

**ausreichen** (28), be sufficient

**ausrüsten** (13), equip

das **Ausschalten** (15), switching-off

**ausschieben** (irr.), push out

**ausschließen** (5) (irr.), shut out

**ausschließlich**, exclusive(ly)

**aussehen** (6) (irr.), look, appear

der **Außendurchmesser** (16), outside diameter

die **Außenluft** (38), outside air

die **Außentemperatur** (36), outside temperature

**außerdem** (22), moreover

**außerordentlich** (22), extraordinary
**äußerst** (23), extreme(ly)
**der Aussteller** (19), exhibitor
**die Ausstellung** (19), exhibition
**aussuchen** (7), seek out
**ausweichen** (21) (irr.), yield, give way
**auswerfen** (25) (Sep. irr.), throw out
**auszeichnen** (28) (Sep.), distinguish
**der Auszug** (14), extract
**der Automat** (7) **(-en)**, automatic machine, automaton
**die Automatisierung** (29), automating

**die Badoberfläche** (31), surface of the bath
**der Bagger** (27), excavator
**die Bahn** (11) **(-en)**, track, path, lane
**2 bahnig** (27), having two lanes
**bald** (5), soon
**barfuß** (5), barefoot
**die Bauart** (16) **(-en)**, construction type
**bauen** (14), build
**das Baugewerbe** (17), building trade
**das Baukastensystem** (29), add-a-plant technique
**baulich** (27), architectural
**der Baum** (5) **(-̈e)**, tree
**die Baumaßnahme** (27), building measure or step
**beachten** (10), study, pay attention to, observe
**beanspruchen** (31), put to the test, strain
**beantworten** (29), answer
**bearbeiten** (3), fashion, work, machine
**die Bearbeitung** (29), treatment
**das Bearbeitungsstadium** (19), treatment stage
**der Bedarf** (19), need, demand
**bedeuten** (27), mean, signify

**bedeutend** (24), significant
**die Bedeutung** (12), significance
**bedienen** (33), operate
**der Bediener** (5), servant, operator
**die Bedienung** (29), working, operation
**der Bedienungshebel** (15), operating lever
**der Bedienungsmann** (33), operator
**das Bedienungspersonal** (17), operating staff
**der Bedienungsstand** (27), state of readiness
**beeinflussen** (17), influence
**beenden** (28), end, finish
**befahrbar** (27), navigable
**befassen** (16) **(sich mit)**, occupy oneself, concern oneself with
**der Befehl** (26) **(-e)**, command
**der Befeuchter** (38), moistener
**die Befeuchtung** (38), moistening
**befinden (sich)** (13), be (situated)
**befindlich** (23), to be found
**die Beförderung** (19), transport
**die Befreiung** (29), liberation
**die Begegnung** (11), meeting together, encounter
**der Beginn** (18), beginning
**beginnen** (32) (irr.), begin
**begrenzen** (31), limit
**der Begriff** (11) **(-e)**, notion, idea, term
**die Begrüßung** (18) **(-en)**, greeting
**behandeln** (12), treat
**bei** (2), with, at, in the case of, on
**das Beispiel** (3) **(-e)**, example
**beispielhaft** (19), typical
**beispiellos** (19), unexampled
**bekannt** (16), known, familiar
**bekanntlicherweise** (36), in a manner well known
**bekommen** (2), get, receive
**die Belagerung** (16), siege
**die Belastung** (16) **(-en)**, load

die **Belastungsimpedanz** (35), load impedance
die **Belegschaft** (30), staff
**beleuchten** (4), light, illuminate
**bemerken** (18), remark, observe
**benötigt** (34), required
**benutzen** (3), use
das **Benzin** (15), petrol
das **Benzinluftgemisch** (14), petrol air mixture
**beobachten** (22), observe
**bepflanzen** (6), plant
**berechenbar**, calculable
das **Berechenbare** (26), the calculable
**berechnen** (26), calculate
die **Berechnung**, calculation
**berechtigt** (23), authorized
der **Bereich** (26) (**-e**), scope, sphere
die **Bereifung** (13), tyres, tyring
**bereit** (39), ready
**bereiten** (31), prepare, cause
**bereits** (12), already
der **Bergbau** (17), mining
**berücksichtigen** (17), take into account
die **Berücksichtigung** (28), regard, consideration
der **Beruf** (8) (**-e**), profession
**berühmt** (30), famous
**berühren** (27), affect
die **Berührung** (11), contact
**beschäftigen** (30), occupy
sich **beschäftigen mit** (17), occupy oneself with
**bescheiden** (21), modest
**beschicken** (28), charge
**beschleunigen**, accelerate
**beschreiben** (28) (irr.), describe
**besichtigen** (19), view, inspect
der **Besitzer** (13), owner
**besonder** (15), special
**besonders** (7), especially
**besprechen** (12) (irr.), discuss

die **Besserung** (20), improvement
der **Bestandteil** (10), component
sich **bestätigen** (19), to be confirmed
**bestehen (aus)** (10, 39) (irr.), consist of, survive
der **Besteller** (29), client
**bestens** (32), very well
**bestimmt** (25), definite, particular
**bestrebt** (17), intent on
der **Besuch** (18) (**-e**), visit
der **Besucher** (19), visitor
der **Betonbedarf** (27), concrete requirements
**betrachten** (12), regard, observe
die **Betrachtung** (17) (**-en**), consideration
**betragen** (13) (irr.), to amount to
**betreffend** (25), in question, concerned
**betreiben** (34) (irr.), drive, power
der **Betrieb** (15) (**-e**), working, operation
**außer Betrieb**, out of action
die **Betriebsanlage** (30) (**-n**), premises
**betriebssicher** (17), reliable, dependable
die **Betriebssicherheit** (34), safety of operation
die **Betriebsweise** (37), mode of operation
das **Bett** (9) (**-en**), bed
**bewährt** (22), reliable
**bewältigen** (19), fulfil
sich **bewegen** (9), move
**bewegend** (5), motive, moving
**beweglich**, mobile, moveable
**bewegt** (22), moved, moving
die **Bewegung** (9) (**-en**), movement
der **Bewegungsablauf** (19), working model
die **Bewegungsaufgabe** (12), problem of movement
**bewirken** (23), effect, bring about

**bewundern** (14), admire
**die Bewunderung** (26), admiration
**die Bezahlung** (2) (**-en**), payment
**die Bezeichnung** (25) (**-en**), term
**beziehungsweise** (27), or, respectively
**sich beziffern** (27), amount
**der Bezirk** (30), district
**bezüglich** (31), relative to
**bieten** (20) (irr.), offer, afford
**das Bild** (28), picture, illustration
**billig** (7), cheap(ly)
**die Bindung** (29) (**-en**), obligation, tie
**bis** (8), up to, till
**bisher** (12), up to now, previously
**blaß** (5), pale
**das Blech** (35) (**-e**), plate
**bleiben** (8) (irr.), stay, remain
**die Bleistiftzeichnung** (10), pencil drawing
**der Blick** (7) (**-e**), aspect, point of view
**der Blickpunkt** (29) (**-e**), point of view
**das Blut** (40), blood
**die Bodenfreiheit** (13), ground clearance
**die Bodennähe** (32), nearness to ground
**der Bogen**, bow
**das Bogendreieck** (23), triangle with curved sides
**der Bohrer** (3), borer, gimlet, drill
**die Bohrmaschine**, drilling machine
**der Bohrtrupp** (19), drilling team
**die Bohrung** (13), bore
**das Böschungspflaster** (27), paving of a slope
**der Bratspieß** (21), roasting spit
**brauchbar** (16), practicable
**brauchen** (3), need
**brausen** (5), roar
**breit** (6), wide, broad

**die Breite** (18), breadth
**die Bremse** (22), brake
**die Bremstrommel** (13), brake drum
**brennen** (28) (irr.), burn
**der Brenner** (31), burner
**der Brennschluß** (32), completion of combustion
**der Brennstoff** (5) (**-e**), fuel
**das Brett** (22) (**-er**), board
**bringen** (13) (irr.), bring
**das Buch** (18), book
**buchen** (17), record, register
**der Buchstabe** (18) (**-n**), letter (of alphabet)
**das Buntmetall** (33), non-ferrous metal
**das Büropersonal** (6), office staff
**der Bürophotograph** (10), office photographer
**bzw.** (= **beziehungsweise**)

**der Chemiker** (7), chemist

**da,** there, as, since (because)
**das Dach** (4) (**-̈er**), roof
**dadurch** (7), by that means
**dagegen** (38), on the other hand
**daher** (20), therefore
**damit** (7), so that, in order that
**das Dämmerlicht** (32), grey light of dawn
**der Dampf** (21) (**-e**), steam, vapour
**der Dampfdruck** (21), steam pressure
**dampfen** (39), steam
**dämpfen** (13), damp, smoothe
**der Dampfkessel** (5), steam boiler
**die Dampfkraft**, steam power
**die Dampfkraftmaschine**, steam engine
**das Dampfkraftwerk,** steam power station

die **Dampfmaschine** (5), steam engine

die **Dampfpfeife** (5), steam siren

der **Dampfstrahl** (21), jet of steam

der **Dampfstrom** (21), current of steam

die **Dampfturbine** (21), steam turbine

**daneben** (5), near by

**dann** (8), then

**darstellen** (11), represent

die **Darstellung** (9) (**-en**), representation

**darüber** (17), over that, over it

**daß** (5) (conj.), that

die **Daten** (n.pl.) (24), data

der **Dauerbetrieb** (33), continuous working

**dauern** (6), last

**davon** (10), of it, from it, of them

**dazu** (21), in addition, furthermore

**decken** (19), cover

die **Deckung** (34), covering, supply

**definieren** (29), define

**dementsprechend** (23), accordingly

**denkbar** (22), thinkable, -bly, surprisingly

**denken** (29) (irr.), think

**denn** (13), for

**derart** (21) of the kind, similar

**derartig** (25), of this sort

**der-die-dasselbe** (17), the same

**deshalb** (12), therefore

**deutsch** (8), German

**d.h.** (= **das heißt**) (29), that is

**dicht** (22), thick, dense

der **Dichter** (18), poet, writer, artist

**dienen** (32), serve

der **Dienst** (11) (**-e**), service

**dimensionieren** (36), proportion

das **Ding** (7) (**-e**), thing

das **Diplom** (8), diploma, degree

**dirigieren** (8), direct

**doch** (32), yet

der **Dom** (40), cathedral

die **Donau** (27), Danube

der **Donnerhall** (40), peal of thunder

**donnern** (40), thunder

die **Doppelkette** (33), twin chain

das **Dorf** (27) (**¨-er**), village

**dort** (1), there

das **Drama** (**-en**), play, drama

die **Drehbank** (9) (**¨-e**), lathe

**drehbar** (16), rotatory

die **Drehbewegung** (23), turning movement

der **Drehdurchmesser** (33), swing

**drehen** (3), turn, rotate

der **Drehkolben** (23), turning piston

die **Drehlänge** (33), turning length

die **Drehmomenteneinheit** (20), torque unit

die **Drehrichtung** (27), direction of rotation

der **Drehsinn** (9), direction of turning

der **Drehstahl** (**¨-e**), lathe tool

das **Dreieck** (10), set square, triangle

**dringen** (16) (irr.), press, urge

**dröhnen** (39), hum, rumble

die **Drosselklappe** (38), throttle valve

der **Druck** (17), pressure

die **Druckgießmaschine** (29), machine for making pressure die castings

die **Druckgußfertigung** (29), manufacture of pressure die castings

der **Druckgußhersteller** (29), maker of pressure die castings

der **Druckknopf** (29) (**¨-e**), press button

die **Druckluft** (20), compressed air

die **Drucklufterzeugung** (17), production of compressed air

das **Druckluftgerät** (20), compressed air equipment

**der Druckluftkompressor** (17), air compressor

**der Druckluftmotor** (20), motor for compressing air

**das Druckluftnetz** (20), compressed air network

**die Druckluftspannung** (17), air pressure

**die Druckschwingung** (12), variation of pressure

**das Druckventil** (12), pressure or safety valve

**dunkel** (32), dark

**durch** (2), through, by

**durchaus** (19), by all means, absolutely

**durchbohren** (3), bore through, pierce

**durchbrechen** (32) (irr.), break through

**durchdrücken** (22), press down

**durchexerzieren** (32), practise

**durchführen** (20), carry out

**die Durchführung** (38), carrying out

**durchhängen** (22), sag

**der Durchmesser** (9), diameter

**durchschauen** (26), see through

**durchsichtig** (10), transparent

**durchwegs** (37), without exception

**die Düse** (21) (-n), nozzle

**dutzendmal** (32), dozen times

**eben** (13), level, even

**ebenfalls** (21), also, likewise

**ebenso** (15), just as

**die Ecke** (23), corner, angle

**der Effekt** (14) (-e), effect

**ehe** (32), before

**eigen** (6), own

**eigenartig** (23), curious

**Eigenbedarfnismaschine** (27), special purpose machine

**der Eigenschein** (40), own radiance

**eigentlich** (26), proper, properly speaking

**eignen (sich)** (16), to be suitable, fit

**einander,** each other, one another

**die Einarbeitung** (12) (**in**), introduction (to)

**der Einbau** (20), installation, addition

**einbauen** (28), build into

**einbüssen** (32) (Sep.), lose, forfeit

**eindeutig** (29), unambiguously

**eindrucksvoll** (19), impressive

**einerseits . . . andererseits** (11), on the one hand . . . on the other

**einfach** (17), simple

**der Einfluß** (36) (¨-e), influence

**die Einfuhr** (30) (-en), import

**einführen** (12), import, introduce

**die Einführung** (-en), introduction

**der Eingang** (20), entrance, entry, introduction

**eingeben** (25) (irr.), put in, feed into

**eingefahren** (11), well-worn, conventional

**eingehen** (20) (irr.), go into

**der Eingußtrichter** (28), funnel

**einhalten** (38) (irr.), check, restrain, maintain

**das Einhalten** (28), keeping to, observance

**die Einhaltung** (38), retention, maintenance

**die Einheit** (11) (-en), unit

**einige** (6), some

**einjagen** (29), inspire (fear)

**die Einleitung** (38), introduction

**einmalig** (19), unique

**einnehmen,** take, receive

**einpassen** (28), fit into

**der Einphasentransformator** (35), single-phase transformer

**einrichten** (13), furnish, equip, arrange

**die Einrichtung** (29), arrangement

der **Einsatz** (20), installation
**einschalten** (15), engage
**einschließen** (30), include
**einsetzen** (30), accomplish, install, employ
die **Einsicht** (11) (**-en**), point of view, insight
**Einspindel-** (30), single-spindle
die **Einspritzung** (17), injection
**einstampfen** (28), ram
**einstellen** (16), terminate, regulate
**einstufig** (21), single stage
**eintreten** (6), enter, come about
**einwandfrei** (20), unobjectionable, flawless
**einwirken** (36) (**auf**), react, operate (upon)
die **Einzelbestellung** (30), individual requirement
das **Einzelfach** (11), one's own speciality
die **Einzelheit** (10), detail
**einzeln** (11, 13), particular, independent(ly), individual
das **Einzelstück** (28), single part
der **Einzelteil** (30), single part
**einzig** (33), single
**einzigartig** (30), unique
das **Eisen** (31), iron
die **Eisenbahn** (40), railway
die **Eisenbahnschiene** (28), rail of railway
der **Eisenkern** (35), iron core, track
der **Eisenreiter** (40), iron horseman
die **Eisenstadt** (40), town of iron
**elektrisch** (4), electric
das **Elektrizitätsnetz** (35), electricity grid
das **Elektronengehirn** (18), electronic brain
**empfangen** (2) (irr.), receive
**empor** (40), upwards
das **Ende** (8) (**-n**), end
die **Energieart** (34), form of energy

die **Energiequelle** (7), source of energy
die **Energieversorgung** (35), supply of energy
**eng** (24), narrow, limited, close
der **Engländer** (9), Englishman
die **Enkelschar** (40), host of grandchildren
**entdecken** (10), discover
die **Entfaltung** (11), unfolding, development
die **Entfernung** (13), distance
der **Entfeuchter** (38), moisture eliminator
die **Entfeuchtung** (39), elimination of moisture
die **Entgratung** (29), deburring, fettling
**enthalten** (13), contain, hold
das **Entladungsgefäß**, discharge vessel
**entleeren** (27), empty
das **Entleerungsbauwerk** (27), emptying structure
**entnehmen** (27) (irr.), remove, take away
**entscheidend** (20), considerable, decided
**entsprechen** (23) (irr.), correspond
**entsprechend** (28), corresponding, suitable
**entstehen** (24) (irr.), come into being, result, arise
**enttäuschen** (13), disappoint
**entweder . . . oder** (6), either . . . or
**entwickeln** (10), develop
die **Entwicklung** (7), development
die **Entwicklungsmöglichkeit** (14), the possibility of development
der **Entwicklungsstand** (14), stage of development
der **Entwurf** (10), sketch
**entzwei** (3), in two

**der Erdball** (40), world
**die Erde** (19), earth
**erdenken** (24) (irr.), think out, invent
**das Erdgas** (34), natural gas
**das Erdöl** (19 (-e), petroleum
**der Erdölvorrat** (19), petroleum reserve
**die Erdwärme** (34), heat of the earth
**erfassen** (26), grasp, comprehend, grip
**erfinden** (7) (irr.), invent
**der Erfinder** (10), inventor
**erfinderisch** (9), inventive
**die Erfindung** (9), invention
**der Erfolg** (11), success
**erforderlich** (21), requisite, available
**erfordern** (27), require
**erforschen** (32), explore
**die Erforschung** (19), research
**erfüllen** (31), fulfil
**die Ergänzung** (11), supplement, completing
**ergeben** (25) (irr.), yield
**sich ergeben** (30), result, arise
**das Ergebnis** (26) (-nisse), result
**erhalten** (2) (irr.), receive, maintain
**sich erheben** (29), rise, occur
**erheblich** (31), considerable
**erhellen** (39), light up
**erhöhen** (16), raise, enhance
**sich erinnern** (40) (**an**), remember
**erkennen** (24) (irr.), recognize, perceive
**erlauben** (27), permit
**die Erlaubnis** (27) (-se), permission
**erledigen** (32), execute, complete
**ermöglichen** (11), make possible
**erneut** (21), anew, once more
**ernstlich** (18), seriously, gravely
**erörtern** (19), discuss
**erproben** (19), test
**errechnen** (27), calculate
**erregen** (14), excite, arouse

**erreichbar** (34), attainable
**erreichen** (28), reach, attain
**erscheinen** (16) (irr.), appear
**die Erschließung** (19), opening up, development
**erschmelzen** (28), melt
**erschöpfen** (7), exhaust
**ersetzen** (7), substitute, replace
**ersinnen** (26), think out, devise
**erst** (9), first, only
**die Erstarrung** (11), stiffening, numbness
**erstens** (13), firstly
**erstrecken** (27), stretch, extend
**erteilen** (16), accord, grant
**ertragen** (7) (irr.), bear, tolerate
**sich erübrigen** (36), to be unnecessary
**erwähnen**, mention
**erwarten** (7), await, expect
**erzeugen** (7), produce, generate
**das Erzeugnis** (30), product
**die Erzeugung** (20), production
**erzielbar** (34), attainable
**erzielen** (9), attain, achieve
**die Erzielung** (21), attaining
**die Esse** (39), chimney, flue
**essen** (6) (irr.), eat
**etwa** (12), perhaps
**etwas** (2), something, somewhat, rather
**ewig** (39), eternal
**exakt** (9), with exactitude
**die Explosion** (32), explosion
**explosionsartig** (14), explosively
**explosionsgefährdet** (20), liable to explosion
**die Extremwerte** (36), outside, maximum value
**exzentrisch** (23), eccentric (ally)

**die Fabrik** (1) (-en), factory
**die Fabrikhallenheizung** (36), heating of vast factory space

das **Fach** (11) (**-er**), subject, speciality
die **Fachdisziplin** (11), specialist or subject discipline
der **Fachkreis** (**-e**), circle of experts
der **Fachspezialist** (11), subject specialist
der **Fachverband** (38), group of experts
der **Fachvortrag** (19), lecture
die **Fachwissenschaft** (11), special branch of study
**fachwissenschaftlich** (11), specialist
**fahrbereit** (15), ready for the road
die **Fahreigenschaft** (22), travelling quality
**fahren** (7) (irr.), travel, go, work
der **Fahrer** (22), driver
die **Fahrt** (15), journey, trip
das **Fahrzeug** (7) (**-e**), vehicle
der **Fall** (17) (**-̈e**), case
**falls** (31), in case
**fast** (14), almost
das **Federbein** (13), telescopic leg or strut
**fein** (14), fine, elegant
der **Felsaushub** (27), amount of rock
die **Fensterachse** (38), window axis
**ferner** (31), moreover
der **Fernlaster** (22), long-distance lorry
**fertigen** (30), produce
die **Fertigung** (17), production
der **Fertigungsablauf** (29), course of production
die **Fertigungsgut** (29), manufactured component
die **Fertigungstechnik** (29), production technique
**fest** (21), firm, solid
die **Festlegung** (27), establishing, fixing
**feststehend** (21), fixed, immobile
**feststellen** (20), determine, discover, establish

die **Feuchtigkeit** (38), dampness, humidity
**feuerbeständig** (28), fire resistant
der **Feuerofen** (40), fiery furnace
**fiebern** (39), to be feverish
der **Fiedelbogen** (9), fiddle bow
der **Filialbetrieb** (30), subsidiary business
die **Filiale** (30), subsidiary
das **Finanzamt** (13), Treasury
**finden** (9), find
die **Firma** (6) (**Firmen**), firm
**flach** (33), flat, even
die **Flamme** (5) (**-n**), flame
**fliegen** (7), fly
das **Fließband** (20), assembly line
**fließen** (27), flow
der **Flug** (32) (**-̈e**), flight
der **Flugkörper** (32), flying object
das **Flugzeug** (7) (**-e**), aeroplane
der **Fluß** (35), river, flux
**flüssig** (28), fluid, liquid
das **Flußkraftwerk** (27), river power station
die **Folge** (20, 24), (**-n**), consequence, sequence
**folgen** (21), follow
**fordern** (31), require
die **Forderung** (17) (**-en**), demand
die **Form** (9) (**-n**), form, mould
der **Former** (28), moulder
die **Formgrube** (28) (**-n**), moulding pit
die **Formherstellung** (28), moulding
**formstabil** (22), rigid
die **Forschung** (19), research
**fortdauernd** (30), continual(ly), continuing
**fortschreitend** (13), progressing, progressive
der **Fortschritt** (26), progress
**fortsetzen** (32), continue
die **Fortsetzung** (23), continuation
die **Frage** (11) (**-n**), question

**der Fräser** (30), milling cutter
**die Frau** (1) (-en), woman, wife
**die Fregatte** (32), frigate
**frei** (15), free(ly)
**das Freie** (13), the open air
**die Freileitung** (35), overhead wires
**freiwerdend** (23), becoming free
**fremd**, strange, foreign
**die Fremdsprache** (25), foreign language
**die Frequenz** (34), frequency
**frisch** (6), fresh
**früh** (6), early, a.m.
**frühzeitig** (31), early
**fühlbar** (37), perceptible
**sich fühlen** (22), feel
**führen** (11), lead
**die Fülle** (19), abundance
**füllen** (19), fill
**die Füllmenge** (27), total
**das Füllungsbauwerk** (27), filling structure
**der Funke** (39) (-n), spark
**das Funktionieren** (20), functioning
**für** (1), for
**der Fußboden** (28), floor
**der Fußgashebel** (22), accelerator pedal
**das Futter** (33), chuck
**der Futterautomat** (30), chucking-automatic

**der Gang** (15) (-̈e), gear
**ganz** (7), quite, whole
**gar** (25), even, at all
**das Gas** (15) (-e), gas
**das Gasdrehgriff** (15), twist grip
**die Gasdurchläßigkeit** (28), permeability to gas
**gasförmig** (34), gaseous
**das Gasgemisch** (23), fuel mixture
**der Gasschieber** (15), fuel control valve

**das Gebäude** (4), building
**die Gebäudeaußenfläche** (36), outside surface of building
**geben** (11), give
**es gibt** (6), there is, there are
**das Gebiet** (12), province, sphere, area
**die Gebläsekühlung** (13), fan cooling
**gebräuchlich** (17), usual, customary
**gebunden** (34), combined
**der Gedanke** (14) (-n), thought
**die Gefahr** (11) (-en), danger
**das Gefäß** (15), vessel
**gefettet** (15), enriched
**gegebenenfalls** (20), if occasion arises
**gegen** (24), against, towards
**gegeneinander** (33), opposite each other
**der Gegensatz** (-̈e), contrast
**die Gegenwart** (7), present
**der Gehalt** (-e), content
**das Gehäuse** (21), housing, casing
**gehen** (4) (irr.), go
**das Gehirn** (26), brain
**gehören** (6), belong
**der Geist** (18), mind, intelligence
**geistig** (24), mental, brain-
**das Gelände** (19), land, ground
**gelangen** (32), reach, arrive at
**das Geld** (2) (-er), money
**die Gelegenheit** (19), opportunity
**der Gelehrte** (26), scientist, savant
**das Gelenk** (22) (-e), link, hinge
**gelingen** (16) (irr.), succeed
**es gelingt mir** (16), I succeed
**gelten** (irr.), be valid
**gelten als** (16), to be regarded as
**gelten für** (20), apply to
**die Gemeinkosten** (17), general expenses
**das Gemisch** (14), mixture
**genau** (28), exact(ly)

die **Genauigkeit** (33), exactitude, accuracy

**Generatorklemmen** (27), generator terminals

**genügen,** suffice

**genügend** (6), enough, satisfying

**gerade** (21), straight, just

das **Gerät** (19) (**-e**), apparatus

das **Geräusch** (**-e**), noise, sound

**geräuschlos** (14), noiseless

**gerecht werdend** (16), satisfying

**gericht,** just

**gering** (16), slight, little

das **Gerüst** (32), scaffolding

**gesamt** (20), total, complete

der **Gesamtanschlußwert** (37), total effective capacity

der **Gesamtüberblick** (19), overall view

der **Gesang** (39) (**-̈e**), song

**geschachtelt** (35), packed

**geschehen** (17) (irr.), happen

die **Geschichte** (9) (**-n**), story, history

die **Geschicklichkeit** (14), skill

**geschickt** (10), clever, dexterous

**geschmolzen** (31), molten

das **Geschütz** (14) (**-e**), gun, cannon

die **Geschwindigkeit** (21) (**-en**), speed, velocity

das **Gesetz** (18) (**-e**), law

das **Gesicht** (5) (**-er**), face

die **Gestalt** (28) (**-en**), figure, form

**gestalten** (16), form, shape

die **Gestaltung** (29), design

**gestatten** (27), allow

**gestern** (5), yesterday

**gesund** (6), sound, healthy

**getrennt** (28), separated, separate(ly)

das **Getriebe** (15), gear box

das **Getriebeöl** (15), gear oil

**gewährleisten** (28), guarantee

**gewaltig** (24), powerful

das **Gewerbe** (34), trade, industry

das **Gewicht** (20), weight

die **Gewichtsverteilung** (22), weight distribution

das **Gewinde** (9), thread

die **Gewindeschneideinrichtung** (33), screw-cutting or threading device

**gewinnen** (18) (irr.), win, gain, obtain

die **Gewinnung** (19), production, obtaining

**gewiß** (9), certain

die **Gewohnheit** (32), custom

**gewöhnlich** (2), usual(ly)

**gewünscht** (15), desired

die **Gezeit** (34) (**-en**), tide

das **Gichtgas** (34), blast furnace gas

**gierig** (5), greedy

die **Gießerei** (20) (**-en**), foundry

der **Gießvorgang** (28) (**-̈e**), pouring process

das **Glas** (4), glass

**glatt** (28), smooth

**glauben** (19), believe

**gleich** (2), equal to, like, the same (**sogleich,** immediately)

**gleichartig** (29), of the same sort

**gleichbleibend** (13), remaining the same, steady

die **Gleichdruckturbine** (21), impulse turbine

**gleichlautend** (25), having the same meaning

**gleichmäßig** (23), regular, uniform

die **Gleichmäßigkeit** (31), uniformity

der **Gleichrichter,** rectifier

**gleichseitig** (23), equilateral

die **Gleichspannung,** direct current

der **Gleichstrom,** direct current

**gleichzeitig** (15), simultaneous(ly)

das **Gleitlager,** sliding bearing

das **Glück** (5), happiness, fortune

**glücklich,** happy (**-ily**), fortunate

**glühen** (5), glow

die **Glut** (40), blaze
der **Gott** (40) (¨er), god
das **Grabrelief** (9), relief on a tomb
der **Grad** (16), degree
die **Granate** (14) (-n), shell
**grau** (5), grey
**graugesprenkelt** (32), sprinkled with grey
die **Grenze** (14) (-n), border, margin, limit
**grenzen** (an), border (on)
der **Grenzwert** (38), marginal value
der **Griff** (15), grip
das **Griffstück** (15) (¨e), grip
**groß** (1), large, big, great
die **Größe** (28), size
die **Größenordnung** (37), order of size
der **Großstadtverkehr** (22), city traffic
**größtenteils** (4), for the most part
die **Grube** (28), pit
der **Grund** (14) (¨e), basis, reason
die **Grundbewegung** (23), fundamental movement
**gründen** (11), found, establish
der **Gründer** (16), founder
die **Grundlage** (12) (-n), foundation, base
**grundlegend** (16), fundamental
der **Grundriß** (35) (-e), outline, sketch
**grundsätzlich** (11), fundamental
die **Grundvoraussetzung** (29), basic condition
**gültig** (16), valid
das **Gummitorsionselement** (22), rubber torsion element
**günstig** (22), favourable
der **Guß** (28) (¨e), pouring, casting process
das **Gußeisen** (31), cast iron
das **Gußstück** (29), casting
der **Gußteil** (29), casting

**gut** (4), good, well

**ha**—abbr. for (**das**) **Hektar** (30), hectare (2½ acres)
das **Haar** (32), hair
die **Haftpflichtversicherung** (13), third party insurance
das **Halbleiterventil**, semi-conductor valve
**halbnackt** (5), half naked
**halten** (3) (irr.), hold, keep
der **Hammer** (3) (¨), hammer
**hämmern** (3), hammer
die **Hand** (5) (¨e), hand
der **Handarbeiter** (8), manual worker
die **Handbremse** (13), hand brake
die **Handbremshebel** (15), hand-brake lever
der **Handel**, trade, commerce
**handeln** (37), concern
die **Handelsorganisation** (30), trading organization
die **Handhabung** (37), handling, operation
die **Handsteuerung** (37), manual control
der **Handwerker** (2), craftsman
**handwerklich** (28), craftsmanlike
**hängen** (5), hang
**hängend** (28), hanging
**hantieren** (32), gesticulate
**harren** (19) (+ gen.), await
**hart** (23), hard
der **Haufen** (5), heap, pile
**häufen** (5), heap up, pile
**häufig**, frequently
der **Hauptmann** (32), captain
der **Hauptnachteil** (20), chief drawback
**hauptsächlich** (17), chiefly
der **Hauptstrom** (27), main stream
die **Hauptverwaltung** (30), central administration

das **Hauptwerk** (27), principal construction

das **Haus** (5) (-er), house

der **Haushalt** (34), household

der **Hebel** (15), lever

**heben** (39), heave

die **Hebung** (27), raising

**hegen** (14), cherish

**heiß** (6), hot

**heißen** (6), be called

**heiter**, gay, bright(ly)

die **Heizanlage** (36), heating plant

**heizbar** (31), capable of being heated

**heizen**, heat

das **Heizöl** (34) (-e), fuel oil

die **Heizung** (36), heating

**hell** (6), bright, light

**hemmen** (16), hinder, hamper

**herabsetzen** (17), reduce

**heranziehen** (27), draw upon

**herausdrücken** (15), push out

**herausfliegen** (14) (irr.), fly out

der **Herausgeber** (27), publisher

**herbeiführen** (11), bring (up, along, forward)

der **Herr** (29) (-en), gentleman, master

**herstellen** (7), produce

die **Herstellung** (28), production

**herunterdrücken** (15), press down

**heute** (6), today

**hierbei** (27), as a result

**hierzu** (15), then

die **Hilfe** (18), help, aid

die **Hilfsanlage** (38), auxiliary plant

das **Hilfsmittel** (16), remedy, expedient, aid

der **Himmel** (5), sky, heaven

**hindernisreich** (27), full of obstacles

**hindurchschlängeln** (22), snake through

**hinten** (13) (adv.), behind

**hinter** (5) (prep. & adj.), behind

das **Hinterrad** (15) (-er), rear wheel

der **Hintersteven** (28), stern post

**hin und her** (23), to and fro

**hinwegtäuschen**, divert, conceal, hide

das **Hirn** (39), brain

**hoch, hoh-** (5), high

die **Hochdruckanlage** (37), high-pressure plant

der **Hochdruckteil** (21), high-pressure section

die **Hochgeschwindigkeitsbohreinrichtung** (33), high-speed drilling device

**hochgewachsen** (32), tall

**höchst** (7), highest, extremely

**hochwertig** (28), high grade, valuable

**hoffen** (32), hope

die **Höhe** (32), height

**holen** (40), fetch, draw

**höllenhaft** (40), hellish

das **Holz** (1) (-er), wood

**hölzern** (9), wooden

das **Holzmodell** (28), wooden pattern

die **Holzwalze** (16), wooden roller

**hören** (5), hear

das **Horizont** (11), horizon

der **Hub** (13) (-e), stroke (of piston)

der **Hubkolben** (23), piston

der **Hubkolbenmotor** (14), piston engine

der **Hubschrauber**, helicopter

der **Huf** (40), hoof

**hydraulisch** (12), hydraulic

die **Hydrodynamik** (12), hydrodynamics

die **Hydrostatik** (12), hydrostatics

**i. allg.** (= **im allgemeinen**) (37), in general

die **Idee** (14) (-n), idea

**ihr** (2) (poss. adj.), her, its, their

**immer** (7), always, ever
**imponieren** (22), impress
**imponierend** (32), imposing
**die Industrialisierung,** industrialization
**die Industrie** (20), industry
**der Industriezweig** (17), branch of industry
**induziert** (35), induced
**infolge** (20), as a result of
**die Informationschau** (19), information show
**der Ingenieur** (2), engineer
**das Ingenieurwesen** (7), engineering
**der Inhalt,** contents
**inhomogen** (31), not homogeneous
**die Innenkontur** (23), inside contour
**die Innenkühlung** (17), interior cooling
**das Innere** (21), inside, interior
**innerhalb** (6) (with gen.), inside
**der Insasse** (22), occupant
**insbesondere** (29), in particular
**insgesamt** (19), in all
**die Instandhaltung** (29), upkeep
**die Instandsetzung** (29), repairing
**intensiv** (20), intensive
**das Interesse** (6), interest
**interessieren** (6), interest
**inzwischen,** meanwhile
**irgend ein** (19), some (form) or other
**irgendwo** (19), somewhere or other
**isoliert** (35), isolated, insulated
**der Italiener** (21), Italian

**das Jahr** (5) (-e), year
**das Jahresarbeitsvermögen** (27), annual work capacity
**das Jahrhundert** (7), century
**die Jahrhundertwende** (16), turn of the century
**das Jahrzehnt** (19), decade
**jahrzehntelang** (17), for decades
**je** (27), each

**je . . . desto,** the more . . . the more
**jeder, jede, jedes** (2), each, every
**jederzeit** (22), at all times
**jedoch** (25), however
**jemals** (19), ever
**jenseits** (26), beyond
**jetzig** (7), present, existing
**jeweilig** (37), existing
**jeweils** (27), every time
**jung** (8), young
**der Junge** (8) (-n), boy

**der Kahn** (27), boat
**der Kaiser** (16), emperor
**kalkulatorisch** (29), accounting
**kalt** (15), cold
**die Kammer** (23, 27) (-n), chamber, basin
**die Karosserieaufhängung** (22), the support of the body work
**kaufen** (19), buy
**kaum** (19), scarcely
**die Kavitation** (28), pitting
**der Keil** (3) (-e), wedge
**kein** (2), no, not a
**keinerlei** (20), no sort of
**kennen** (19), know, be acquainted with
**die Kenntnis** (17), knowledge
**der Kern** (28), core, nucleus
**der Kernkasten** (28), core box
**der Kernmacher** (28), core maker
**die Kernmarke** (28) (-n), core-print
**der Kerntrockenofen** (28), core drying oven
**die Kesselschaltung** (37), range of boilers
**der Kesseltyp** (37), type of boiler
**die Kette** (27) (-n), chain
**das Kind** (5) (-er), child
**das Kinderspiel** (22), child's play
**klaffen** (40), yawn
**die Klappe** (38) (-n), valve ("flap")

**klar** (29), clear
**klassisch** (14), classical
**der Klebestreifen** (15), adhesive strip
**das Kleid** (5) (-er), dress, clothes
**kleiden** (6), dress
**klein** (23), small
**der Kleinwagen** (13), small car
**klettern** (32), climb
**die Klimaanlage** (38), air-conditioning plant
**die Klimatisierung** (38), air conditioning, making an artificial climate
**die Klippe** (27), rock
**klirren** (39), clang
**der Klub** (6), club
**knacken** (23), crack
**kochen** (40), boil
**die Kohle** (5) (-n), coal
**der Kohlenstoffstahl** (28), carbon steel
**der Kohlenwasserstoff** (34), hydrocarbon
**das Koksofengas** (34), coke oven gas
**der Kolben** (14), piston
**der Kolbenkompressor** (17), piston compressor
**die Kolbenmaschine** (21), piston engine
**der Kolbenweg** (14), path of the piston, travel
**kommen** (5), come
**kompliziert** (7), complicated
**können** (4), be able
**konstant** (9), constant
**die Konstellation** (36), configuration
**der Konstrukteur** (10), designer
**das Konstruktionsbüro** (10), drawing office
**konventionell** (14), conventional
**konzentrieren** (30), concentrate
**der Körper** (5), body
**körperlich** (18), physical
**korrespondieren** (26), correspond

**die Korrosionsgefahr** (37), danger of corrosion
**die Kosten** (18) (pl.), costs
**die Kosteneinsparung** (20), economy
**die Kraft** (5) (-e), power, force
**der Kraftfluß** (14) (-e), flow of power
**kräftig** (33), strong, powerful
**die Kraftlinie** (35) (-n), line of force
**der Kraftstoff** (15), fuel
**der Kraftstoffhahn** (15), fuel cock
**das Kraftwerk** (34), power station
**krank**, sick, ill
**die Krankheit** (2), illness
**der Kreis** (39) (-e), circle, circuit
**der Kreiskolbenmotor** (23), rotating piston engine or motor
**die Kubatur** (27), cubic quantity
**das Kugellager** (16), ball bearing
**kulturell** (6), cultural
**der Kunde** (29) (-n), customer
**die Kunst** (10), art, skill
**der Kunsttischler** (9), cabinet maker
**der Kupferdraht** (35), copper wire
**kuppeln** (38), link, couple
**die Kupplung** (15), clutch
**der Kupplungshebel** (15), clutch lever
**die Kurbelwelle** (22), crankshaft
**kurz** (6), short(ly)
**kurzgeschoren** (32), cut short
**der Kurzschlußknopf** (15), short-circuiting switch button
**die Kybernetik** (24), cybernetics

**die Lage** (29) (-n), position
**die Lagerart** (16), kind of bearing
**die Lagerung**, support, bearing
**die Lampe** (19), lamp
**das Land** (9) (-er), land, country
**lang** (5), long
**die Länge** (18), length
**langsam** (7), slow(ly)
**der Längsdrehanschlag** (33), length stop

die **Längsdreharbeit** (33), longitu-
dinal traversing
der **Lärm** (1), noise
**lassen** (9) (irr.), let, allow, cause
der **Lauf** (13), course, movement
**laufen** (5) (irr.), run
der **Laufkranz** (21), rotating ring
das **Laufrad** (21), rotor
die **Laufruhe** (21), smoothness of
running
**laut** (1), loud(ly), aloud
**leben,** live
**lebendig** (10), alive
die **Lebensdauer** (22), life span, life
**leer** (10), empty
das **Leergewicht** (13), weight when
empty
der **Leerlauf** (15), neutral gear
**legieren** (35), alloy
**legiert** (15), alloyed, mixed, com-
pounded
das **Legierungselement** (31), alloy-
ing element
**lehren** (26), teach
die **Lehrzeit** (10), apprenticeship
period
**leicht** (6), light, easy (-ily)
die **Leichtgängigkeit** (22), ease of
movement
**leider** (20), unfortunately
**leisten** (11), achieve
die **Leistung** (2), achievement, per-
formance
**leistungsfähig** (7), efficient, produc-
tive
die **Leistungsfähigkeit** (17), efficiency
**leistungsmäßig** (27), as regards out-
put
**leistungsstark** (22), strong in per-
formance
der **Leistungswert** (37), capacity,
effectiveness
**leiten** (21), direct, lead

der **Leitkranz** (21), set of redirecting
fixed blades
der **Leitungsverlust** (17), delivery
or transmission loss
**lenken** (11), guide, direct
der **Lenker** (13), linkage, stay,
handlebars, link
die **Lenkung** (22), steering
**lernen** (10), learn
**letzt** (30), last, recent
die **Leuchtölgewinnung** (19), making of lamp oils
die **Leute** (5), people
**licht** (27), clear, lucid
das **Licht** (32) (-er), light
die **Lichtpause** (10), photographic
print
das **Lied** (39) (-er), song
die **Liefermenge** (17), delivery
**liefern** (5), deliver
**liegen** (5) (irr.), lie
das **Lineal** (3), rule(r)
**links** (15), on the left
die **Lochkarte** (24), punched card
der **Lochstreifen** (24), punched
strip
**locken** (40), entice
**logisch** (26), logical
**lohen** (39), blaze
der **Lohn** (2) (-̈e), payment, wages
**lösen** (3, 7), loosen, slacken, solve
**loslassen** (15), let go
die **Loslösung** (29), separation,
liberation
die **Lösung** (12) (-en), solution
die **Lücke** (22), gap
die **Luft** (38), air
die **Luftaufbereitung** (38), air puri-
fication
die **Luftbehandlung** (38), air treatment
das **Luftloch** (15), air hole
die **Luftmenge** (17), quantity of air
**lullen** (40), lull

das **Lunkermittel** (28), feeder compound

**machen** (1), make, do
**mächtig** (32), mighty
**magazinlos** (29), without a magazine
die **Magazinmaschine** (29), magazine loading machine
die **Magnetisierbarkeit** (35), magnetic property
die **Mahlzeit** (6), meal
**man** (4), one
**mancher** (23), many-a
der **Mann** (1) (⁻er), man
**mannigfaltig** (38), manifold
die **Manövrierfähigkeit** (32), manoeuvrability
der **Marineflieger** (32), naval airman
die **Marke** (15) (-n), mark
das **Markenbenzin** (15), proprietary petrol
die **Maschine** (1) (-n), machine
**maschinell** (29), mechanical(ly)
der **Maschinenbau** (12), machine construction
der **Maschinensatz** (27), machine set
der **Maschinenzubehörteil** (30), machine component
das **Maß** (12) (-e), measure, dimension
die **Masse** (-n), mass, bulk
**maßgebend** (17), decisive(ly)
**mäßig** (18), moderate
die **Maßnahme** (20) (-n), measure
der **Maßstab** (28) (-e), standard
**maßstabgerecht** (10), scale, true to scale
das **Material** (9) (-ien), material
die **Mathematik** (8), mathematics
**mathematisch** (7), mathematical(ly)
die **Mauer** (27) (-n), wall
die **Mauernflanke** (40), wall flank
der **Mechaniker** (1), mechanic
**mechanisch** (18), mechanical
**mehr** (2), more

**mehrere** (16), several
**Mehrfachstähle** (33) (pl.), multiple tools
der **Mehrspindelstangenautomat** (30), multi-spindle bar automatic
der **Meißel** (3), chisel
**meist** (29), most, usually
der **Meisterzeichner** (10), master or expert draughtsman
die **Menge** (29), quantity
der **Mensch** (6) (-en), person, human being
**menschlich** (26), human
die **Messe** (19) (-n), fair
die **Meßeinrichtung** (29), measure device
**messen** (3), measure
das **Metall** (1) (-e), metal
das **Metallgewerbe** (17), metal industry
der **Metallurge** (7), metallurgist
die **Methode** (11) (-n), method
der **Militärflugplatz** (32), military aerodrome
**mindestens** (8), at least
das **Mindestmaß** (32), minimum
die **Minute** (14) (-n), minute
der **Mischer** (31), mixer
die **Mischkammer** (38), mixing chamber
**mitnehmen** (13) (irr.), take with
der **Mittag**, midday
die **Mittagspause** (6), midday break, lunch hour
das **Mittel** (29), means
der **Mitteldruckteil** (21), medium pressure section
**Mitteleuropa** (27), central Europe
der **Mittelpunkt** (29), centre
**mittels** (3) (+ gen.), by means of
das **Mittelwasser** (27), mean water-level
**mittler** (35), mean, average
der **Modellaufbau** (19) (-ten), model

**der Modellschreiner** (28), pattern maker

**die Modellschreinerei** (28), pattern maker's shop

**möglich** (14), possible

**die Möglichkeit** (11), possibility

**möglichst** (17), utmost possible

**monatlich** (2), monthly

**die Monatschrift** (14), monthly periodical

**der Motor** (13) (-en), motor, engine

**der Motorblock** (13), engine block

**der Motorlauf** (14), running of the motor

**müde** (5), tired

**der Müll** (34), refuse

**müssen** (4), be obliged, must

**der Musterbestandteil** (10), prototype component

**das Musterprodukt** (19), sample product

**die Mutter** (3) (-n), nut

**der Nabenteil** (28), hub piece

**nach** (6), after, to (a place), according to

**nachahmen** (10), imitate

**nachdem** (10), after

**nachgießen** (28), add (by pouring)

**der Nachkühler** (20), after-cooler

**die Nachricht** (23) (-en), news

**das Nachschlagewerk** (12), work of reference

**das Nachschwärzen** (28), blackening

**die Nacht** (4) (-̈e), night

**der Nachteil** (14), disadvantage

**nachträglich** (20), subsequent

**nachvollziehen** (26), achieve afterwards

**die Nadel** (32) (-n), needle

**nah, näher, nächst** (7), near, nearer, nearest or next

**nahezu** (7), almost, practically

**namens** (14), by name

**die Natur** (34), nature

**die Naturwissenschaft** (8), natural science

**neben** (6), near, by

**das Nebenprodukt** (19), by-product

**nehmen** (15), take

**neigen** (33), bend, incline

**der Nenndurchmesser** (33), nominal diameter

**die Nervenzelle** (26), nerve cell

**nett** (6), neat, nice

**neu** (7, 8), new, modern

**von neuem** (11), anew

**der Neuaufbau** (27), rebuilding

**die Neuerung** (27), innovation

**das Neuron** (26), neurone

**nicht** (1), not

**nichts** (26), nothing

**nichtstationär** (12), non-stationary, unsteady

**nie**, never

**der Niederdruckteil** (21), low pressure section

**niedrig** (6), low

**niemals** (22), never

**noch** (6), yet, still

**nochmals** (15), again, once more

**der Nocken** (33), cam

**die Nockensteuerung** (33), cam control

**normalerweise** (13), normally

**der Normalverbrauch** (13), normal consumption

**der Normbauteil** (12), standard component

**der Normverbrauch** (13), normal consumption

**die Note** (8), mark

**nötig** (9), necessary

**notwendig** (10), necessary, essential

**die Notwendigkeit** (30), need

**nun** (25), now

**nur** (8), only

**die Nuß** (23), nut

nützen (19), utilize
die Nutzlänge (27), effective length
die Nutzlast (27), maximum load
nützlich (9), useful
der Nutzungsfaktor (27), utilization factor

ob (10), whether
oben (15), up, above (nach oben, upwards)
die Oberfläche (28), surface
der Oberhafen (27), upper harbour or dock
oberhalb (27), above
die Oberspannung (35), high voltage
obgleich (9), although
obwohl (19), although
das Ofenfutter (31), furnace lining
offen (5), open (adj.)
öffnen (15), open (verb)
oft (10), often
ohne (6), without
das Ohr (4) (-en), ear
das Öl (15) (-e), oil
ölbeheizt (31), oil heated
die Öldruckbremse (13), oil pressure brake
die Öleinfüllschraube (15), oil filling cap
das Ölgesellschaft (19), oil company
ölglatt (14), oil-smooth
die Ölhydraulik (12), oil hydraulics
die Oper (6) (-n), opera
die Ordnung (20), order
der Ort (27), village
Österreich (27), Austria
oxydierend (31), oxidizing

ein paar (33), a few
paarweise (27), in pairs
das Papier (10), paper
der Parkplatz (6) (-̈e), car park
die Paßstraße(22), mountain pass

das Patent (16), patent
die Pause (10), tracing
die Pendelachse (13), swing axle
das Pendelkugellager (16), self-aligning ball bearing
der Petrochemiker (19), petroleum chemist
die Pfanne (28), pan
der Pfeil (35) (-e), arrow
die Pferdestärke (13), horse-power
die Photographie (6), photography, photograph
die Physik (8), physics
physikalisch (12), physical
der Physiker (7), physicist
der Pilot (32) (-en), pilot
der Plan (8) (-̈e), plan
planen (27), plan
das Platzverhältnis (21), space condition
der Pleuel (22), connecting rod
die Pleuelstange (14), connecting rod
praktisch (8), practical
die Praxis (37), practice
präzise (22), precise(ly), accurate(ly)
die Präzision (9), exactitude
der Preis (34), price, cost
primitiv (5, 9), simple, very early
das Prinzip (26), principle
probieren (15), test, try
das Problem (7) (-e), problem
der Problemkomplex (11), maze of problems
die Produktionsstufe (30), stage of production
das Profil (30), section
das Profildrehen (33), form turning
das Profilgerät (33), former
das Projekt (18), project
die Projektion (10), projection
prüfen (15), test
der Prüfstand (15), testing stand

die **Prüfung** (28) (-en), examination
**PS** = **Pferdestärke** (13)
das **Pulver** (14), powder
die **Pumpe** (12) (-n), pump
der **Punkt** (22), point, item

die **Quelle** (7) (-n), source
die **Querdreharbeit** (33), cross traversing
**quergelenkt** (13), cross-braced
der **Querschlitten** (33), cross slide
der **Querschnitt** (-e), cross section
die **Querschnittzeichnung** (23), cross-sectional drawing
der **Querträger** (22), crosspiece, traverse support

das **Rad** (9) (-̈er), wheel
das **Räderrasseln** (39), rattle of wheels
der **Radstand** (13), wheel base
der **Rahmen** (13), frame
das **Raketentriebwerk** (32), rocket motor
der **Rand** (32), edge
**rasch** (12), quick
die **Rationalisierung** (18), rationalization
der **Rauch** (6), smoke
**rauchen** (32), smoke
das **Rauchgas** (37) (-e), smoke gas
die **Rauchwolke** (5), smoke cloud
der **Raum** (23) (**Räume**), space
die **Raumbedingung** (38), space condition
**räumen** (33), broach
die **Raumheizung** (36), space heating
der **Räumhub** (33), stroke
die **Räummaschine** (30), broaching machine
die **Räumnadel** (33), broach
die **Raumtemperatur** (36), temperature of the room

das **Räumwerkzeug** (30), broaching tool
die **Rechenanlage** (24), calculating machine
der **Rechenautomat** (24), computer
das **Rechengerät** (26), computer
die **Rechenmaschine** (7), computer, calculating machine
**rechnen** (17), calculate
**recht** (32), right
**rechts** (15), on the right
die **Regel** (25), rule
die **Regelfrage** (37), problem of regulation
die **Regelgröße** (36), regulating factor
die **Regelung** (36), regulation, adjustment
das **Regelventil** (12), bleeder valve
der **Regelverstärker** (38), automatic volume control
die **Reibahle** (30), reamer
die **Reibung** (16), friction
**reichen** (19), reach, last
der **Reifendruck** (15), tyre pressure
die **Reihe** (12), row, series
**rein** (15), clean, pure
die **Reinhaltung** (20), keeping clean
die **Reinheit** (20), purity
die **Reise** (18) (-n), journey
das **Reißbrett** (10), drawing board
**reißen** (40) (irr.), tear, snatch
die **Reißfeder** (10) (-n), drawing pen
die **Reißschiene** (10) (-n), tee square
das **Rennöl** (15), racing oil
die **Reparatur** (20) (-n), repair
das **Repetitionsdrehen** (33), repetition turning
das **Reservat** (26), preserve, reserve
**resp.** (= **respecktiv**) (27), respectively
das **Resultat** (26), result
**revolutionär** (14), revolutionary

der **Revolver-automat** (30), turret screw automatic
die **Revolverdrehbank** (30), turret lathe
der **Revolverschlitten** (33), turret slide
der **Rhythmus** (29), rhythm
**richten** (7), direct
**richtig** (22), correct, exact
die **Richtung** (14) (-en), direction
die **Riemenscheibe** (4), belt-pulley, driving wheel
**riesig** (27), gigantic
das **Rohr** (14) (-e), tube, barrel (of gun)
die **Rohre** (-n), tube, radio valve
der **Rohrrahmen** (22), tubular framework
die **Rohzeichnung** (10), rough sketch
die **Rolle** (31), role, part
das **Rollenlager** (16), roller bearing
der **Rollwiderstand**, rotation resistance
**rot** (5), red
**rotieren** (9), rotate
**routinemäßig** (26), routine
die **Rückluft** (38), return air
die **Rücksicht** (25), consideration, regard
der **Rückstau** (27), impounded water
**rückwärts** (13), backwards
das **Ruder** (28), rudder
das **Ruderhorn** (28), rudder post, rudder-horn
**ruhen** (16), rest
**ruhig** (15), peaceful(ly), gently
die **Rundschau** (36), review
**rundum** (23), round about

die **Säge** (3), saw
**sammeln** (31), collect
der **Samstag** (5), Saturday
**sämtlich** (18), all

die **Sättigung** (38), saturation
**sauber** (6), clean
**sausen** (39), rush
**sausend** (39), rushing(ly)
**schaffen** (13) (irr.), create, develop
der **Schall** (32), sound
die **Schallmauer** (32), sound barrier
der **Schaltdrehgriff** (15), twist-grip for gear changing
das **Schalten** (22), gear-changing, switching
der **Schaltgriff** (15), gear-change grip
der **Schalthebel** (15), gear lever
der **Schaltplan** (12), wiring diagram, circuit diagram
der **Schaltschrank** (29), switchboard
die **Schalttechnik** (20), switching technique
die **Schaltung** (35), connexion
die **Schaltvorrichtung** (33), indexing fixture
die **Schamottemasse** (28), fire-clay mass
**schätzen** (19), estimate
die **Schätzung** (19), estimate
die **Schaufel** (21), shovel, blade
der **Schaufelkranz** (21), ring of blades (turbine)
**schaufeln** (5), shovel
das **Schaufelrad** (21), paddle wheel
die **Scheibe** (4) (-n), disc
**scheinbar** (32), apparent
**scheinen** (4) (irr.), shine, appear
der **Scheinwerfer** (15), head lamp
**scheuen** (20), avoid
**schieben** (16) (irr.), push, shove
das **Schiff** (21) (-e), ship
die **Schiffahrt** (27), navigation
der **Schiffaufbau** (28), ship construction
der **Schiffbau**, shipbuilding
die **Schlacke** (31), slag, dross
die **Schläfe** (39), temple (of head)

der **Schlag** (39), blow
**schlagen** (3) (irr.), strike, hit
das **Schlagwerkzeug** (3), striking tool
das **Schlagwort** (29), catchword
die **Schlange** (40), serpent
die **Schleifbank** (4) (**-e**), grinding machine
die **Schleifmaschine** (30), grinder
das **Schleppschiff** (27), tug
der **Schleppzug** (27), train of barges
die **Schleuse**, lock, sluice
die **Schleusenanlage** (27), system of locks
die **Schleusenkammer** (27), lock basin
die **Schleusenkammersohle** (27), bottom of lock basin
das **Schleusenunterhaupt** (27), sluice outlet
**schließen** (21), close
der **Schlitten** (33), slide
**schlingen** (9) (irr.), sling, wind
der **Schlot** (5) (**-e**), chimney
**schmackhaft** (6), tasty
**schmal** (21), thin, narrow
die **Schmelze** (28), melt
der **Schmelzofen** (31), blast furnace
der **Schmelzrest** (31), pig metal, what is left after the melt
das **Schmiedestück** (33), forging
**schmutzig** (5), dirty
die **Schneidegeschwindigkeit** (33), cutting speed
**schneiden** (3) (irr.), cut
**schnell** (2), quick(ly)
der **Schnitt** (35) (**-e**), section
die **Schnittzeichnung** (12), sectional drawing
die **Schnur** (9) (**-en**), cord
**schnurren** (39), whirr
**schon** (7), already
**schön** (6), fine, beautiful
**schonen**, protect

**schonend** (6), protective
**schöpferisch** (26), creative(ly)
der **Schornsteinhügel** (40), hill of factory chimneys
der **Schoß** (40), bosom
die **Schräglage** (16), sloping position
die **Schraube** (1) (**-n**), screw
die **Schraubenfeder** (22), spiral spring
die **Schraubenform** (9), screw form
der **Schraubenschlüssel** (3), spanner
der **Schraubenzieher** (3), screw driver
die **Schraublehre** (3), micrometer
der **Schraubstock** (3) (**-e**), vice
**schreiben** (irr.), write
der **Schreiber** (6), clerk
das **Schreibgerät** (29), recording apparatus
**schreien** (1) (irr.), shout
**schrill** (40), shrill(y)
der **Schub** (23), thrust, push, impulse
die **Schubleistung** (32), thrust
die **Schule** (8) (**-n**), school
der **Schüler** (18), scholar
**schützen**, protect
die **Schwäche** (10), weakness
**schwanken** (38), sway, oscillate
die **Schwankung** (34), fluctuation
der **Schwarm** (32), swarm
**schwarz** (5), black
der **Schwede** (21), Swede
**schweißbar** (25), weldable
**schwenken** (15), turn
**schwer** (16), heavy
der **Schwerpunkt** (22), centre of gravity
**schwierig** (12), difficult
die **Schwierigkeit** (31), difficulty
das **Schwindmaß** (28), amount of shrinkage
die **Schwingachse** (22), independent axle
das **Schwungrad** (9), fly wheel

**sehen** (4) (irr.), see
**sehr** (1), very
**die Seilzugbremse** (13), cable brake
**sein** (1) (irr.) (verb), be
**sein** (2), his, its
**seit** (6) (prep.), since
**seitdem** (10) (conj.), since
**die Seite** (29), side
**der Sekundenzeiger** (32), second hand
**selber** (17), itself
**selbst** (19), oneself, myself, himself, herself, etc., even
**selbsttätig** (16), automatic
**selten** (19), rare, seldom
**semantisch** (25), semantic
**setzen** (21), put, set
**sicher** (22), safe, secure
**die Sicherheit** (17), safety
**sichern** (28), secure, make safe
**die Sicherung** (28), insurance, securing
**sichtig** (36), clear
**das Silizium** (35), silicon
**sinnlos** (26), meaningless(ly)
**sinnreich** (23), ingenious
**sinnvoll** (26), meaningful(ly)
**der Sklave** (16), slave
**sobald** (25), as soon as
**sofort** (24), immediately
**sogar** (38), even
**sogenannt** (15), so-called
**solch** (14), such
**solide** (22), solidly
**der Sommer** (5), summer
**der Sonderdruck** (36), off-print, reprint
**das Sonderheft** (37), special number
**die Sondermaschine** (30), special-purpose machine
**sondern** (8), but (after a negative)
**die Sonne** (4), sun
**die Sonneneinstrahlung** (36), sun's rays

**die Sonnenenergie** (34), energy of the sun
**das Sonnenlicht** (4), sunlight
**sonst** (15), otherwise
**sonstig** (34), other
**sorgfältig** (10), careful(ly)
**soweit** (12), in so far as
**sowie** (12), as well as
**sowohl . . . als auch** (20), both . . . and
**sowohl . . . wie auch**
**der Spalt** (40) (-e), split
**die Spannung** (7), stress, tension, voltage
**der Spannungsabfall,** voltage drop
**die Spannvorrichtung** (33), chuck
**der Spannzylinder** (20), chuck
**das Spant** (28), rib
**sparsam** (22), economical
**spät** (5), late
**der Spatz** (13), sparrow
**das Speicherwerk** (25), storage mechanism, "memory"
**speien** (40) (irr.), spit
**der Speisesaal** (6)(-säle), dining room
**spenden** (39), bestow
**sperrig** (28), bulky
**die Spezialisierung** (11), specialization
**spielen** (15), play
**die Spindel** (9) (-n), spindle
**der Spindelstock** (9), headstock
**der Spiralbohrer** (30), twist drill
**die Spitze** (13, 33), maximum, centre, pivot
**die Spitzenleistung** (30), first-class achievement
**spitzenlos** (30), centreless
**der Sportplatz** (6)(-e), sports ground
**die Sprache** (8), language
**sprechen** (1), speak
**sprühen** (39), spray, sprinkle
**die Spurweite** (13), track width

die **Staatskunst** (18), politics
**stabil** (13), sturdy, steady
die **Stadt** (40), town
der **Stahl** (4), steel
der **Stahlguß** (28), cast steel
die **Stahlmenge** (28), mass of steel
der **Stahlschaufel** (21), steel blade
**stammend** (27) (**aus**), dating (from)
**stampfen** (5), stamp, pound
das **Standbild** (16), statue
**ständig** (13), constant(ly)
**stark** (1), strong, robust
die **Startanlage** (13), starter mechanism
**statisch** (35), static
**stattfinden** (34), take place
**stattfindend** (35), taking place
die **Stauhöhe** (27), height of dammed water
der **Stauraum** (27), reservoir of impounded water
die **Staustufe** (27), barrage level
**stehen** (13) (irr.), stand
**steigen** (19), climb, increase
**steigern** (17), increase
die **Steigung** (22), gradient
**steil** (22), steep
die **Steinbearbeitung** (17), stone-cutting
der **Steinblock** (16) (⸚e), block of stone
die **Steingewinnung** (17), stone-quarrying
der **Steinwurf** (27), shifting of stone
**stellen** (17), place, set
**stellen** (**sich**) (16), pose itself
**steuern** (24), control
der **Steuerschieber** (12), control valve
die **Steuerung** (24), control
die **Steuerungslehre** (24), study of control mechanisms
die **Steuerungstechnik** (20), technique of control

der **Steven** (28), stern, stern post
die **Stimme** (4) (**-n**), voice
der **Stoff,** material
die **Störgröße** (36), factor of disturbance
der **Stoßdämpfer** (13), shock-absorber
die **Straße** (5) (**-n**), road, street
die **Straßenlage** (22), road-holding quality
die **Strecke** (27), stretch, distance, length
der **Streufluß** (35) (**-e**), eddy current
der **Strom** (20), river, current, electric current
der **Stromabschnitt** (27), section of river
**strömen** (5), stream, flow
die **Stromerzeugung** (34), production of current
**stromlinienförmig** (32), streamlined
der **Stromregler** (12), flow regulator
der **Stromrichter,** transformer, converter
die **Stromrichtertechnik,** transformer technology
**stromseits** (27), on the stream side
die **Strömungsmaschine** (17), flow engine
die **Strömungsrichtung** (21), current direction
der **Strömungsvorgang** (12), flow phenomenon
das **Stück** (28), piece, part
das **Stückgewicht** (28), weight of the piece (casting)
die **Stückzahl** (29), number of pieces
der **Student** (8) (**-en**), student
die **Studienzeit** (8), period of study
**studieren** (8), study
der **Studierende,** student
das **Studium** (8) (**-ien**), study

die **Stufe** (32), step, stage
**stufenweise** (21), in stages
**stumm** (32), mute(ly)
die **Stummheit** (26), dumbness
die **Stunde** (5) (**-n**), hour
**Stundenkilometer** (32) (pl.), kilometres per hour
der **Stundenruf** (40), call of the hours
**stürzen** (32), fall
**stützen** (39), prop, support
die **Stützmauer** (27), supporting wall
die **Suche** (19), search
**suchen** (14), seek, try
die **Sulfitlauge** (34), sulphite liquor
die **Summe** (2), sum
**summen,** hum
das **Summen** (6), humming
**surren** (39), hum

die **Tabelle** (12), table, list
die **Tachonadel** (22), speedometer needle
die **Tafeldarstellung** (19), diagrammatic representation
**täglich** (6), daily
**tanken** (15), put into tank
der **Tanz** (40) (**-̈e**), dance
der **Taupunkt** (38), dew point
die **Taupunktunterschreitung** (37), control of the dew point
der **Techniker** (2), technician
**technisch** (8), technical, technological
der **Teil** (10) (**-e**), part
**zum Teil** (36), partly
**teilen,** share, divide
die **Teilung** (28), division
das **Tempo** (7), speed, pace
die **Testflugreihe** (32), series of test flights
**teuer** (20), dear, expensive
das **Theaterstück** (6) (**-e**), play

die **Thermodynamik** (12), thermodynamics
**tief** (22), deep(ly), low
die **Tinte** (10), ink
das **Tor** (5) (**-e**), gate
der **Torf** (34), peat
**tragen** (11) (irr.), carry, bear, wear
der **Träger** (28), bearer
die **Tragfähigkeit** (16), carrying capacity
die **Tragfläche** (32), wing
die **Transferbearbeitungsstraße** (29), flow production line
der **Transformator** (35) (**-en**), transformer
die **Transformatorschaltung,** transformer switching
**treiben** (6) (irr.), drive, perform, engage in
**trennen** (15), separate
der **Trennpfeiler** (27), dividing pier, or pillar
der **Treppelweg** (27), towing path
das **Trichtergewicht** (28), funnel weight
die **Trickdarstellung** (19), trick representation, cartoon
die **Trockenheit** (20), dryness
**trocknen** (28), dry
**trommelförmig** (23), drum-shaped
**trostlos** (5), cheerless, bleak
**trotz** (26) (+ gen.), in spite (of)
das **Tuch** (10) (**-̈er**), cloth
**tun** (32), do
**tupfen** (15), make use of the "Tupfer," "tickle"
der **Tupfer** (15), carburettor operating rod, "tickler"
die **Tür** (5) (**-en**), door
die **Turbinenwelle** (21), turbine shaft

**u.a.** (**unter anderen**) (19), among other things, *inter alia*

**über** (5), over, above, about
**überall** (20), everywhere
der **Überblick** (27), survey, review
die **Überdruckturbine** (21), high-pressure turbine
die **Übereinstimmung** (28), conformity
**überflüssig** (7), superfluous
**überführen** (30), transfer
**überfüllen** (5), overfill
der **Übergang** (29) (ӟe), transition
**überhaupt** (35), after all
die **Überhitzung** (31), over-heating
das **Überholen** (22), overtaking
das **Überkleid**, overall
die **Überlegung** (23), consideration
**überprüfen** (20), test
die **Überschallgeschwindigkeit** (32), supersonic speed
**überschauen** (19), survey
**überschreien** (4) (irr.), shout above
**übersetzen** (25), translate
der **Übersetzer** (18), translator
die **Übersetzungsroboter** (24), translating robot
das **Übersetzungsverhältnis** (22), transition ratio
**übersichtlich** (27), clear, distinct, easily
**überstauen** (27), cover with water
**übertragen** (13) (irr.), transfer, transmit
die **Übertragung** (34), transmission
die **Überwachungseinrichtung** (29), inspection device
**überzeugen** (18), convince
**üblich** (14), usual
**üblicherweise** (17), usually
die **Übung** (8), exercise, practice
das **Ufer** (27), bank (of river)
die **Uferlänge** (27), length of bank
die **Uhr** (5) (-en), clock, watch
**um** (6) (prep.), round, at
**um . . . zu** (3), in order to

die **Umdrehung** (23), rotation
**umfangreich** (27), extensive
**umfassen** (30), embrace, include
**umfassend** (27), comprehensive
**umgeben** (18) (irr.), surround(ed)
**umgekehrt** (11), opposite, reverse, vice versa
**umgewandelt** (25), transformed
die **Umluft** (38), surrounding air
der **Umluftanteil** (38), share of the surrounding air
**umschlingen** (35) (irr.), wind round
das **Umschmelzen** (31), re-casting
**umstalten** (24), transform
der **Umstand** (36) (ӟe), circumstance
die **Umstellung** (29), change, conversion
**umsteuerbar** (21), reversible
**umwaldet** (40), surrounded by forest
die **Umwälzung** (7), revolution
**umwandeln** (25), transform
die **Umwandlung** (34), transformation
**unabhängig** (22), independent
**unangenehm** (20), unpleasant
**unbedingt** (22), definite, decided, absolute
das **Unbekannte** (32), the unknown
**unberechenbar** (26), incalculable
**unbestreitbar** (20), incontestible
**undenkbar** (9), unthinkable
**uneingeschränkt** (26), unrestrained
**unentbehrlich**, indispensable
**ungebildet** (18), uneducated
**ungefähr** (27), about
das **Ungetüm** (22), monster
**ungezählt** (39), unnumbered
**unheimlich** (26), uncanny
die **Universität** (8), university
**unkontrollierbar** (31), uncontrollable
**unlegiert** (28), unalloyed, pure
**unmittelbar** (31), direct
**unmöglich**, impossible

unregelmäßig (33), irregular
unser (7), our
unten (15), down, below
unter (5), under
unterbringen (27) (irr.), house
unterentwickelt (9), backward
der Unterhafen (27), lower harbour
  or dock
unterlegen, p.p. of unterliegen
untern (38), lower
unternehmen (19) (irr.), undertake
der Unternehmer (38), employer
unterschreiten (38) (irr.), go below
die Unterspannung (35), low voltage
unterstützen (6), support
die Untersuchung (16), investiga-
  tion, examination
die Untiefe (27), shallow, shoal
ununterbrochen (9), uninterrupted
unverändert (9), unchanged
unwahrscheinlich (20), improbable
die Unzahl (24), endless number
UpM (= Umdrehungen in der
  Minute) (33), r.p.m.
uralt, primitive, ancient
ursprünglich (10), original(ly)
utopisch (18), Utopian

der Vater (Väter) (40), father
v. Chr. (= vor Christus) (16), B.C.
das Ventil (22), valve
veränderlich, variable
verändernd (36), variable(ly)
die Veränderung (38), alteration,
  change
verankert (28), anchored
die Verantwortung (29), responsi-
  bility
verarbeiten, process, work up
die Verarbeitung (19), processing
das Verarbeitungsstadium (19),
  manufacturing stage
verbauen (28), build up

verbessern (18), improve, correct
die Verbesserung (17), improvement
verbinden (26), combine
die Verbindung (24), combination
der Verbrauch (19), use, consumption
verbrauchen (28), consume, use up
der Verbraucher (19), user, con-
  sumer
die Verbreitung (20), spread, extension
verbrennen (19) (irr.), burn up
die Verbrennung (37), combustion
die Verbrennungsmaschine, internal combustion engine
der Verbrennungsmotor (14), internal combustion engine
verbringen (10) (irr.), spend (time)
verbunden (14) (f. verbinden), connected
verdammt (26), condemned
verdichten (23), compress
verdrängen (29), displace
vereinbart (34), combined
vereinfachen (37), simplify
die Verengung (11), limitation,
  restriction
das Verfahren (19), procedure
der Verfasser (20), author
die Verfeinerung (29), improvement
der Verfeuerungsprozeß (37), firing
  process
verfolgen (11), pursue
verfügbar (12), available
die Verfügbarkeit (34), availability
die Verfügung (17), disposition,
  disposal
zur Verfügung stellen (17), to
  make available
die Vergangenheit (7), past
der Vergaser (15), carburettor
vergewissern (15), assure, make sure

**vergießen** (28) (irr.), pour
**vergleichen** (19) (irr.), compare
**vergrößern** (20), increase
**das Verhältnis** (7) (-se), relationship, condition, circumstance
**verhältnismäßig** (21), comparatively
**die Verkaufskraft** (30), sales force
**der Verkehr** (34), traffic, association
**das Verkehrsgewühl** (22), traffic jam
**die Verknüpfung** (35), linkage
**der Verlag** (11), publishing firm
**verlangen** (13), demand
**verlassen** (6) (irr.), leave
**sich verlassen auf** (22), rely upon
**die Verlegung** (27), transfer, removal
**die Verleimung** (28), gluing
**verlieren** (27) (irr.), lose
**verlockend** (23), alluring, attractive
**vermeiden** (31) (irr.), avoid
**vermischen** (15), mix
**vermitteln** (11), mediate
**vermögen** (21) (irr.), be able
**vernünftbegabt** (26), gifted with intelligence
**vernünftig** (29), intelligent
**die Verrichtung** (29), job, duty
**die Verriegelung** (15), locking
**verschaffen** (20), procure, provide
**verschieden** (6), different
**der Verschleiß** (22), wear and tear
**verschleißfest** (22), durable, hardwearing
**verschließen** (15) (irr.), shut
**die Verschmutzung** (20), dirtying, pollution
**verschollen** (40), forgotten, missing
**versehen** (21) (irr.), (**mit**), provide(d) with
**versetzen** (21), set, put
**verständnisbereit** (29), ready to understand
**verstehen** (8) (irr.), understand

**die Versteifung** (28), stiffening
**die Verstrebung** (28), bracing, reinforcing
**versuchen** (24), try
**das Verteilnetz** (37), distribution network
**der Verteilstrang** (37), distribution line
**das Verteilsystem** (38), distribution system
**die Verteilung** (38), distribution
**vertragen** (32) (irr.), bear, tolerate
**vertraut** (32), familiar
**vertreten** (30), represent
**der Vertreter** (24), representative
**die Vertretung** (30), agency
**der Vertrieb** (19), sale, distribution
**die Vertriebsorganisation** (30), distributing organization
**verursachen** (29), cause
**verwenden** (12) (irr.), use, apply, employ
**die Verwendung** (20), use, application
**der Verwendungszweck** (20), purpose
**verwerten** (34), utilize
**verwirklichen** (16), realize, achieve
**die Verwirklichung** (16), realization
**verwunderlich** (17), surprising
**verziert,** adorned
**verzögern,** retard
**viel** (1), much
**viele** (1), many
**vielfach** (37), frequently, in many cases
**vielfältig** (20), manifold, various
**vielleicht** (8), perhaps
**vielrohrig** (14), of many barrels (gun)
**vielseitig** (9), many-sided, versatile
**die Vielzahl** (34), multiplicity
**das Vierganggetriebe** (13), four-speed gear

die Vierscheiben-Lamellenkupp-lung (13), four-disc laminated clutch

vierstöckig (6), of four storeys

voll (2), full, complete

die Vollast (32), full load

vollbesetzt (22), filled

vollbringen (27) (irr.), achieve

die Vollendung (10), completion, perfection

vollgasfest (22), steady at full throttle

völlig (20), fully, completely

vollkommen (24), complete, perfect

vollständig (30), complete(ly)

die Vollständigkeit (19), complete-ness, perfection

volumetrisch (17), volumetric(ally)

von (1), of, from

vor (4), before, in front of

vor allem (37), above all

vorausberechnen (26), calculate in advance

voraussetzen (16), assume, presuppose

die Voraussetzung (20), assumption, condition

vorbehalten (13), reserved

vorbei (19), past

vorbereiten (8), prepare

die Vorbereitung (15), preparation

vorbeugen (37), prevent

die Vorderachse (13), front axle

das Vorderrad (13), front wheel

die Vorderradbremse (15), front wheel brake

der Vorgang (33) (-e), process, phenomenon, event

der Vorgebildete (12), trained person

vorgenommen (22), arranged, planned

vorgeschrieben (28), prescribed

vorgesehen (28), intended

vorhanden (28), available, present

vorher (26), previously

der Vorherd (31), preliminary stove

vorherrschen (17), predominate

vorkommen (9) (irr.), occur, appear

der Vorläufer (16), precursor

vormittags, before noon, a.m.

vorn(e) (15), in front

nach vorn(e) (15), forwards

der Vorort (5), suburb

die Vorrichtung (7), device

vorschieben (32) (irr.), push forward

vorsehen (27) (irr.), provide

sich vorstellen (26), imagine

die Vorstellung (29) (-en), idea

der Vorteil (17) (-e), advantage

vorteilhaft (37), advantageous

vortreiben (irr.), drive forward

der Vortriebe (9) (-e), propulsion

vorwärts (13), forward

vorwiegend (36), predominant(ly)

waagerecht (33), horizontal

das Wachstum (30), growth

der Wagen (6), car

während (9), while, during

wahrnehmen (19), make use of

wahrscheinlich (19), probable

walten (26), rule, hold sway

das Wälzlager (16), roller bearing

die Walzmaschine für Gewinde (30), thread generator

wandern, hike, ramble

die Wärme (39), heat

der Wärmeanfall (36), onset of warmth

der Wärmeaustauscher (38), heat exchanger

wärmeentwickelnd (28), exo-thermic

die Wärmeentwicklung (20), de-velopment of heat

der Wärmeerzeuger (37), heat-producer

**das Wärmekraftwerk** (34), thermal power station

**die Wärmelieferung** (36), supply of heat

**die Wärmeüberführung** (31), heat transfer

**der Warmhalteofen** (31), heat-retentive stove

**warnen** (5), warn

**die Wartung** (20), maintenance

**wartungsfrei** (22), not needing maintenance

**was** (7), what

**waschen** (6) (irr.), wash

**der Waschraum** (6), wash room

**das Wasser** (6), water

**der Wasserdampf** (38), water vapour

**wassergekühlt** (22), water-cooled

**das Wasserkondensat** (20), condensed water

**die Wassermenge** (27), amount of water

**der Wasserspiegel** (27), water level

**die Wasserstandbeeinflussung** (27), influencing the water-level

**der Webstuhl** (24), loom

**der Wechsel** (14), change, alteration

**der Wechselrichter,** inverted converter, mutator

**die Wechselspannung** (35), alternating voltage

**der Wechselstrom,** alternating current

**wechselweise** (23), alternately

**der Weg** (11) (-e), way, road, path

**die Wehranlage** (27), dam

**das Wehrfeld** (27), space between the piers

**der Wehrpfeiler** (27), barrage pier

**weil** (6), because

**die Weise** (11), manner

**die Weisheit** (18), wisdom

**weit** (11), far, distant, extensive

**die Weite** (27), **distance, width,** extent

**weitertreten** (15) (irr.), paddle forwards

**weitgehend** (31), extensive

**weitverzweigt** (26), widely ramified

**welcher, -e, -es** (29), which

**die Welle** (16) (-n), shaft

**die Welt** (18), world

**die Weltausstellung** (34), world fair or exhibition

**weltbekannt** (16), world famous

**der Weltraum** (32), space

**der Wendekreis** (13), turning circle

**wenden** (12) (irr.), turn, direct

**wendig** (22), easily steered, manageable

**wenig** (4), little

**weniger** (6), less, minus

**wenn** (2), if, when, whenever

**die Werbeabteilung** (27), advertising department

**werden** (4) (irr.), become. Auxiliary used to form Future and Passive.

**werfen** (7) (irr.), throw, cast

**das Werk** (15) (-e), work, factory, works

**der Werkführer** (1), supervisor

**der Werkmeister** (1), foreman

**die Werkstätte** (1) (-n), workshop

**der Werkstoff** (28), material

**das Werkstück** (3) (-e), workpiece, job

**das Werkzeug** (3) (-e), tool

**die Werkzeugmaschine** (20), machine tool

**die Werkzeugmaschinenbauabteilung** (30), section for the construction of machine tools

**der Werkzeugschlitten** (9), tool slide

**der Wert** (-e), value

**wertlos** (19), worthless

wertsteigernd (19), with the object of increasing the value

wertvoll (12), valuable

wesentlich (10), essential(ly)

wichtig (12), important

die Wicklung (35) (-en), winding

wie (5), how, as, like

wieder (30), again

wiedergeben (21), give back

wiederholen (14), repeat

wiederkehren (24), recur

die Wiederverwendung (31), re-use

wiegen (30), rock

wiegen (28) (irr.), weigh

die Wiese (40) (-n), meadow

wild (27), wild

der Windanfall (36), onset of wind

das Winkeldrehen (33), taper turning

der Winkelmesser (10), protractor

der Winter (4), winter

wirken (13), operate

wirklich (21), real

die Wirklichkeit (26), reality

wirksam (26), effective

der Wirkungsgrad (17), operating standard, degree of efficiency

wirtschaftlich (17), economical(ly)

die Wirtschaftlichkeit (31), economy, saving

der Wirtschaftsweg (27), service road

die Wissenschaft (11), knowledge, science

der Wissenschaftler (19), scientist, savant

wissenschaftlich (16), scientific

das Wissenschaftsgebiet (11), sphere or department of science

wo (8), where

wobei (38), by which

die Woche (2), week

wöchentlich (2), weekly

wohl (16), well, probably

sich wölben (40), arch over

wollen (6), be willing, want

das Wort, Wörter (18), word, words

das Wörterbuch (25), dictionary

sich wundern (18), wonder

der Wunsch (16) (¨e), wish

die Würde (26), dignity

die Wüste (32), desert

die Zähigkeit (15), viscosity

die Zahl (25) (-en), number

zählen (36), count

der Zahlenwert (25), numerical value

zahllos (5), innumerable

zahlreich (12), numerous

die Zahnlenkung, rack-steering

die Zahnstange (22), rack

die Zahnstangenlenkung (13), rack-steering

das Zehnfache (19), ten times

zeichnen (28), draw

der Zeichner (6), draughtsman

die Zeichnung (10) (-en), drawing

zeichnungsgerecht (28), in accordance with the drawing

zeigen (10), show

die Zeit (9) (-en), time

zur Zeit (34), at the moment

das Zeitalter (32), age

zeitlich (29), periodic, temporal

der Zeitpunkt (34), moment

der Zeitraum (29), period of time

die Zeitschrift (11) (-en), periodical

der Zellenverdichter (17), "cell"-compressor

zentral (9), centrally

die Zentrale (34), power-station

die Zentralrohr-Konstruktion (13), central tube construction

zerstören (28), destroy

ziehen (15) (irr.), pull, draw

**ziehen (fest)** (3), to tighten
**das Ziel** (11) (**-e**), aim, goal, target
**ziemlich** (38), fairly, rather
**die Zierkappe** (13) (**-n**), hub cover
**das Zierstück** (9), decorative piece or component
**das Zimmer** (18), room
**der Zirkel** (10), pair of compasses
**zögern**, hesitate
**zu** (22), to, at, too
**das Zubehör** (19), accessories
**die Zubringeeinrichtung** (29), feed device
**zudecken** (13), cover
**zuerst** (13), first
**zufällig** (19), by chance
**der Zuflußkanal** (27), filling-channel
**zufolge** (21) (prep. with Gen.), in consequence (of)
**zufügen** (21), add
**das Zuführmagazingerät** (30), automatic feed attachment
**die Zugabe** (31), addition
**zügeln** (40), bridle, curb
**die Zugspindel** (9), feed screw
**zukommen** (25) (irr.), fit, apply
**die Zukunft** (7), future
**zukünftig** (8), future
**zuleiten** (27), return, lead back
**zuletzt** (19), finally
**die Zulieferindustrie** (19), supply industry
**zumachen** (5) (Sep.), close, shut
**zunächst** (16), at first, first of all
**zünden** (14), ignite
**der Zünder** (14), igniter, detonator
**die Zündkerze** (14) (**-en**), spark-plug
**die Zündung** (15), ignition
**zunehmen** (12) (Sep. irr.), increase
**zunehmend** (20), increasing
**zuordnen** (25), co-ordinate, link, relate

**zurück** (12), back
**zurückführen** (14), lead back
**zurücklenken** (11), turn back, lead back
**den Blick zurücklenken** (11), turn the attention back
**zurückverwandelt** (25), changed back
**zusammen**, together
**die Zusammenarbeit** (11), co-operation
**zusammengedrückt** (23), pressed together
**zusammenfassen** (27), comprise, include
**der Zusammenhang** (36) (**⸚e**), connexion
**das Zusammenketten** (30), linking
**zusammenpassen** (28), fit together
**die Zusammensetzung** (31), composition
**die Zusammenstellung** (12), summary, list
**der Zusatz** (15) (**⸚e**), additive
**die Zusatzeinrichtung** (30), attachment
**zutreffen** (31), be convenient
**zwangsläufig** (11), inevitable
**zwar** (26), admittedly, it is true
**und zwar** (27), and moreover
**der Zweck** (11) (**-e**), purpose, aim
**zweier** (27), of two
**der Zweitakter** (13), two-stroke (engine)
**das Zweitakt-Gemisch** (15), two-stroke mixture
**zweitrangig** (14), second-rate
**der Zwerg** (32), dwarf
**der Zwischenboden** (27), subsoil
**der Zylinder** (9), cylinder
**das Zylinderrollenlager** (16), parallel roller bearing

# INDEX

*The numbers refer to pages*

195